AF658403

CAPACITY DEVELOPMENT

Concepts and Methods

THE AUTHORS

Dr. C. Satapathy retired as Head and Dean of Extension Education from Orissa University of Agriculture and Technology (OUAT) in the year 2002. At present he is working as Director of Amity Humanity Foundation at Bhubaneswar, Odisha under the umbrella of Amity University, Noida. He obtained Ph.D. degree from Indian Institute of Argil Research in the year 1981. He has teaching and research experience of 34 years. He has guided more than 30 Ph.D. Scholars, authored 14 books in various disciplines in addition to publishing more than 200 research papers in National and International journals.

Prof. Santosh Kumar Rout, Dean Institute of Agricultural Sciences SOADU graduated in Agriculture Extension Education, OUAT in 1978. He did Ph.D. in Agricultural Extension Education, OUAT in addition to present position he has worked in OUAT in the capacity of Dean of Extension Education. He has been conferred with FAO fellowship and Best Extension Scientist and published number of papers in National and International Journals.

Mr. Dipak K. Sahu working as Research Associate of Amity Humanity Foundation, Bhubaneswar with 14 years of experience in various areas of livelihood system. In addition he has worked in water management, forest welfare, displacement of tribal population, land based enterprise and other areas of social importance. He has expertise in designing field research program, data collection, analysis and report writing of various natures. He has developed special interest to work with rural people for development activities, adoption of farm technologies and training.

CAPACITY DEVELOPMENT

Concepts and Methods

– Authors –

Dr. C. Satapathy

Dr. Santosh K. Rout

Mr. Dipak K. Sahu

2026

Associated Publishing Company

A Division of

Astral International Pvt. Ltd.

New Delhi – 110 002

© 2026 AUTHORS

ISBN 978-93-5919-349-6

Publisher's Note:

Every possible effort has been made to ensure that the information contained in this book is accurate at the time of going to press, and the publisher and author cannot accept responsibility for any errors or omissions, however caused. No responsibility for loss or damage occasioned to any person acting, or refraining from action, as a result of the material in this publication can be accepted by the editor, the publisher or the author. The Publisher is not associated with any product or vendor mentioned in the book. The contents of this work are intended to further general scientific research, understanding and discussion only. Readers should consult with a specialist where appropriate.

Every effort has been made to trace the owners of copyright material used in this book, if any. The author and the publisher will be grateful for any omission brought to their notice for acknowledgement in the future editions of the book.

All Rights reserved under International Copyright Conventions. No part of this publication may be reproduced, stored in a retrieval system, or transmitted in any form or by any means, electronic, mechanical, photocopying, recording or otherwise without the prior written consent of the publisher and the copyright owner.

Published by : **Associated Publishing Company**
A Division of
Astral International Pvt. Ltd.
– ISO 9001:2015 Certified Company –
4736/23, Ground Floor, Ansari Road
Darya Ganj, New Delhi-110 002
Ph. 011-43549197, 8130496929
e-mail: info@astralint.com
Website: www.astralint.com

Printed at : **Replika Press Pvt Ltd**

Preface

The book **'Capacity Development: Concepts and Methods'** written as per Restructured Revised Syllabus prescribed by ICAR for Extension Education students (Ext. 505) of all Agricultural Universities of India. The book has been written as per Block and Units prescribed taking and referring materials from different sources to make reading easy for the students. We have acknowledged the sources and references. We are grateful to the sources mentioned or not mentioned. We hope the book will serve the purpose for which it is written.

Dr. C. Satapathy

Dr. Santosh K. Rout

Mr. Dipak K. Sahu

Contents

Chapter 1

Capacity Development: An Overview

This Chapter Contains

Sl.No.	*Contents*
1.1	Capacity Development - An overview
1.2	Training, Capacity Building, Capacity Development and HRD - Meaning and Differences
1.3	Need and Principles of Capacity Development
1.4	Types and Levels of Capacities
1.5	Organizational Capacities and Individual Capacities (Technical, Functional and Leadership Skills)
1.6	Capacity Building based on Structure
1.7	Capacity Building based on Context

1.1 Capacity Development - An Overview

Definition

Capacity development refers to the approaches, strategies and methodologies used by developing country, and/or external stakeholders, to improve performance at the individual, organizational, network/sector or broader system level.

Why Capacity Development?

Capacity Development is of recent interest as many development projects face the problem of shortage of skilled man power. The short comings have limited the sustainability of many projects in priority areas particularly in eradication of poverty. Long term capacity building process is the need of the time.

Meaning of Capacity Development

Capacity Development (CD) is viewed from two angles. Some authors refer CD as the process or approach to reduce poverty while other angle emphasizes on development of individual or organization to perfectly handle the problems of poverty eradication. The individuals and organizations need to be equipped with abilities, skill, values, attitude, motivation and resources so that goal of poverty eradication can be achieved. Research reveals that individuals and organizations should be competent in mastering these capacities to achieve the social goal of poverty eradication. However, both schools of thought can be combined to achieve the objectives of different projects.

Capacity Development can be viewed from three levels. These are, sector, organization and individual.

1. **The sector/network level:** Many of the nations are now focussing on sectorial level because of heavy investment. The priority areas are policy reform service delivery, strategies and program frame work.
2. **The organizational level:** This capacity level focuses on organizational structures, processes, resources and management issues. Typically the inputs include technical assistance, budgetary or infrastructure support to individual organizations, or support for institutional linkages.
3. **The individual level:** At individual level emphasis is given on skill covering resource management. In agriculture, small marginal and landless farmers lack up to date skill to contribute to production and productivity. It is established that these group of people with required abilities can contribute significantly for which capacity building is being viewed seriously

Capacity Development Strategies

There are various strategies, which may include any one of the following, or potentially a combination of them to achieve the goal of development.

1. Eliminating old or inappropriate capacity
2. Making better use of existing capacity
3. Building up or strengthening existing capacity
4. Providing space for innovation or creative use of innovation
5. Creating new capacity

Objective of Capacity Development

The objective of capacity development is to;

1. Enhance effective use of skills, abilities, and
2. Address issues of values, attitudes, motivations and conditions in order to support sustainable development.

Capacity Development Principles

The capacity Development is based on the following principles.

1. Broad-based participation and a locally driven agenda
2. Building on local capacities
3. Ongoing learning and adaptation
4. Long term investments
5. Integration of activities at various levels to address complex problems

1.2 Training, Capacity Building, Capacity Development and HRD - Meaning and Differences

A. Training

Training is the action of informing or instructing with employees on a certain task in order to help them improve their performance or knowledge. Trained persons can take job to possible highest level.

Effective training will mean the activities to achieve specific outcomes required. In addition, the workers need to gain or maintain the skills and knowledge they need to perform their work, direct others to perform work and to supervise work. Effective training should be cost efficient, ensuring that time and money is a good investment.

Four Stages of Training

1. Pre-employment training
2. Induction training

3. In-service training
4. Career development training

1. **Pre-employment training:** This is when new employee has already has previous education or training. This training has happened prior to new employee being hired. Although this level of training indicates that previous education may be beneficial, this will often not be directly related to the job at hand. Therefore, educational certificates and qualifications should only be used as an initial indicator as to what skills and knowledge an individual already has, but it will need to make individuals fit to work.
2. **Induction training:** This is also known as orientation training. This is a formal way to induct the new employees into organisation. It introduces them to be members of the teams. Induction training will generally occur right upon the new employee commencing their new role. This stage of training can range from one day to several weeks depending on the nature of the job. It is the most important stages, as this is the time frame that new employees will form impressions on the company and decide if they want to stay with organisation.
3. **In-service training:** This stage of training may involve the use of a schedule or checklist. It is highly beneficial to invest time or money at this stage of training, as the new employees will be more motivated, feel like they are of value to organisation. This will occur throughout the period of employment. It is particularly important when the processes or products change.
4. **Career development training:** This is similar to in-service training. It is related to employees being able to be promoted or expanding their knowledge and understanding of the job at hand.

B. Capacity Building

Capacity building is defined as the process of developing and strengthening the skills, instincts, abilities, processes and resources that organizations and communities need to survive, adapt, and thrive in a fast-changing world.

The Four Capacity Building Pillars of Entrepreneurship

Four Pillars	*Description*
1. Operational Capacity Building	It involves working in a variety of business operations for a period of time prior to diving into entrepreneurship. This is where we gain valuable insight into what makes businesses tick.

Four Pillars	*Description*
2. Management Capacity Building	Operational experience is the, gaining management experience in a field or business will be directly applicable to managing own business. The valuable experience we gain managing operations, resources and people will give competency
3. Financial Management Capacity Building	With gained skills, we will need to be able to analyse financial statements, looking at trends and indicators and what those all mean to business.
4. Personal Capacity Building	Demonstrating strong traits and behaviours such as dedication, perseverance, ambition, determination, strong-will, openness, honesty, transparency, fairness, *etc.* may move the pathway to become a successful entrepreneur.

C. Capacity Development

Capacity development is the improvement in an individual's or organization's facility (or capability) "to produce, perform or deploy.

Six Pillars of Capacity Development

Refers to the current situation of each Capacity Pillar that describes what is working/not working in each pillar; the pillars referred to are,

1. Structure
2. Competencies
3. Management systems
4. Enabling policies
5. Knowledge and learning and, lastly
6. Leadership

Capacity development is the process through which individuals, organizations and societies obtain strengthen and maintain the capabilities to set and achieve their own development objectives over time (UNDP, 2009). Capacity development is essential for making a sustainable contribution to addressing violence against women and girls. While often equated with training only, the concept of capacity development entails much broader components and considerations.

D. Human Resource Development

Human resource development refers to an assortment of training programs that help people adjust to their new roles and learn more about the organization and its culture. A protocol for HRD teaches employees skills they may need for their jobs and grants them access to resources that can help them succeed.

Human resource development helps organizations improve employee performance and ability. The main functions include employee training, performance management and sourcing and acquiring the best talent.

HRD stands for Human Resources Development. Human Resource Development (HRD), is a part of human resource management, is in charge of employee training and development. HRD is a framework for assisting people in improving their skills, knowledge, and abilities to increase the performance of a business.

Difference between Capacity Development and HRD

While HRD tends to focus on training and development in organizational contexts (Torraco and Lundgren 2020), CD often considers broader applications, which include institutional building and development at the community and societal levels (Lusthaus, Adrien, and Perstinger 1999).

Difference between Human Resource and Capacity Development

The two concepts differ in their focus, scope, and target outcomes as explained the following section.

Differences Respect to	*Human Resource Development*	*Capacity Development*
1. Focus	Human resource development's main focus is to enhance skills, knowledge, and performance of individuals within organizational context, for improving employee effectiveness and job satisfaction	Capacity development takes a broader perspective that involves individual, organizational, institutional, and systemic dimensions, with aim to strengthen overall capabilities to address complex challenges and achieve sustainable development outcomes
2. Scope	HRD programs often operate within the boundaries of an organization and mainly target individual employees or specific work teams.	Capacity development programs enlarge organizational boundaries end involve a wide range of stakeholders, which include government agencies, civil society organizations, communities, and other actors involved in development efforts.

Differences Respect to	*Human Resource Development*	*Capacity Development*
3. Out-comes	HRD initiatives have the main focus on short to medium term outcomes, like improved employee performance, increased productivity, and enhanced job satisfaction.	On other hand, capacity development programs has the main goal for long-term sustainability and systemic change, with outcomes that involves strengthened institutions, enhanced governance frameworks, and improved socio-economic conditions at the community and societal contexts

1.3 Need and Principles of Capacity Development

A. Need

Common needs for capacity development include public awareness, education; information management policy, legislative and regulatory frameworks; organizational mandates structures; economic and financial sustainability.

The importance of capacity is that it allows the business to fulfil demand levels and provides flexibility to make diverse products. Efficient capacity management ensures that no extra money is spent.

Benefits of Capacity Building

Capacity building is valuable and important because of its many long-term impacts.

1. Capacity building approaches purposefully minimize dependency on outside experts as sources of knowledge, resources, and solutions to community issues.
2. Capacity building fosters a sense of ownership and empowerment, so that community partners gain greater control over their own future development.
3. It Strengthens confidence, skills, knowledge, and resources that increase from capacity building efforts on one project may enhance a community partner's ability to envision and take action on other projects.
4. Capacity building efforts are sensitive to the particularities of local culture and context, and, as a result, often lead to more feasible and appropriate community solutions than approaches that lack a capacity building focus.
5. Capacity building approaches to community work acknowledge that growth, learning, and change occur reciprocally; that is, both community

partner and collaborative community work. Ideally, the community partner will be more effective and successful in addressing community issues.

B. Principles of Capacity Development

The UNDP approach to supporting capacity development is driven by values and consists of a conceptual framework and a methodological approach.

It is based on the following basic principles:

1. The UNDP approach makes the concept of national ownership tangible. It is about the ability to make informed choices and decisions.
2. It addresses power relations, mind sets and behaviour change. It therefore emphasizes the importance of motivation as a driver of change.
3. Capacity development is a long-term process. It can be promoted through a combination of shorter-term results that are driven from the outside and more sustainable, longer-term ones that are driven from the inside.
4. It requires sticking with the process under difficult circumstances.
5. The approach links the enabling environment, as well as organizations and individuals and promotes a comprehensive approach.
6. It looks beyond individual skills and a focus on training to address broader questions of institutional change, leadership, empowerment and public participation.
7. It emphasizes the use of national systems, not just national plans and expertise. It discourages stand-alone project implementation units; if national systems are not strong enough, it deems that those be reformed and strengthened, rather than bypassed.
8. It requires adaptation to local conditions and starts from the specific requirements and performance expectations of the sector or organization it supports. There are no blueprints.
9. It makes the link to broader reforms, such as those in education, wage structures and the civil service. There is little value in designing isolated, one-off initiatives.
10. It results in unplanned consequences that must be kept in mind during the design phase. These should be valued, tracked and evaluated.
11. It measures capacity development systematically, using good-practice indicators, case evidence and analyses of quantitative and qualitative data, to ensure that objective judgements are made about capacity assets and needs, as well as the progress achieved.

1.4 Types and Levels of Capacities

A. Types of Capacity Development

There are two kinds of capacity which are the functional capacity and technical capacity, which has to be developed at all the three levels: According to the UNDP, functional capacity includes:

1. Functional Capacity

(i) **Capacity to engage stakeholders:** It implies motivating and mobilizing stakeholders, building partnerships, and networks, managing diversity, *etc.*

(ii) **Capacity to assess situation and define the goals:** It implies evaluating different information and then developing goals for capacity development.

(iii) **Capacity to formulate policies and strategies:** Once the goals have been set, relevant execution policies and strategies have to be developed. The capacity to manage human and financial resources. Assessment capacity, *i.e.*, accumulation of lessons learnt, collecting feedback and adjusting the policies and strategies accordingly.

2. Technical Capacity

On the contrary, technical capacity relates to specialized knowledge of a particular area or sector, such as climate change, epidemics, legal issues, *etc.*

B. Levels of Capacity Development

According to UNDP, capacity development issues must be addressed at three levels:

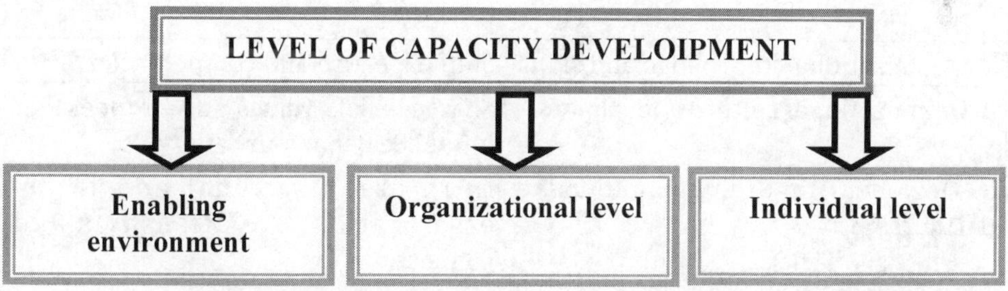

1. **Enabling environment:** The enabling environment relates to issues such as policies, legislation, institutional arrangements, leadership, political processes, power relations and social norms. These factors govern how different parts of society prioritize, operate, and engage each other. If favourable conditions are developed, it will allow effective capacity development.
2. **Organizational level:** It relates to internal policy systems, strategies and arrangements that help any institution to work together and achieve the

goal of capacity development. Since capacity assessment is mostly done at the institutional level, it is important to improve the specific capacities like institutional arrangements, leadership, knowledge and accountability of the institutions working on capacity development.

3. **Individual level:** It relates to the skill and knowledge of individual persons, which can be improved through formal education, training, developing practices and networks in communities, *etc.*

1.5 Organisational Capacities and Individual Capacities (Technical, Functional and Leadership Skill)

A. Organisational Capacities

The organizational level of capacity comprises the internal policies, arrangements, procedures and frameworks that allow an organization to operate and deliver on its mandate, and that enable the coming together of individual capacities to work together and achieve goals.

Five Key Dimensions of Organizational Capacity

The Collaborative has organized different aspects of organizational capacity into five categories or "dimensions." Each of these dimensions has multiple sub dimensions.

Dimensions

Dimensions	*Relates to*
1. Organizational resources	Concrete materials and assets
2. Organizational infrastructure	Organizational structure, protocols, and processes
3. Organizational knowledge and skills	Staff expertise and competencies
4. Organizational culture and climate	Shared beliefs, values, and attitudes that influence behaviour
5. Organizational engagement and partnership	Intra-organizational and inter-organizational relationships and connections

B. Individual Capacities (Technical, Functional and Leadership Skills)

(a) **Technical:** Technical capacity of an organization is the combination of physical, operational ability of staffs and manager's technical knowledge to manage the organization effectively.

In simple terms, when we describe Technical capacity of any organization, it means the physical and operational ability of the organization, including technical knowledge and capability of the staffs and managers to produce desired outputs of the technical capacity *viz.* Technical choices of the organization and Skill acquisition of the staffs.

(b) **Functional:** Functional capability means an individual's ability to perform certain activities taking into consideration factors such as mobility, communication, self-care, self- direction, work tolerance or work skills. Functional capability can be determined on the basis of an evaluation of rehabilitation potential.

Functional capacity refers to an individual's capability, under controlled conditions, to perform tasks and activities that are necessary or desirable in their lives.

(c) **Leadership skills:** Leadership capacity is the ability to think and then act in ways that are more effective during times of increasing VUCA (volatility, uncertainty, complexity, ambiguity) and rapid change. Increasing leadership capacity (rather than just teaching competencies) is known as vertical development.

Top 15 Essential Leadership Qualities to Thrive in Today's Workplace (with Tips)

Leadership Qualities	*Tips for Enhancement*
1. Integrity	1. Demonstrate the values and behaviours 2. Take accountability 3. Ensure transparency 4. Implement policies 5. Reward and recognition 6. Utilize feedback
2. Honesty, Ethical behaviour	1. Offer accurate information 2. Encourage open dialogue 3. Provide truthful feedback 4. Preserve confidentiality
3. Fairness: Equality within the team	1. Clear and transparent policies 2. Maintain standards 3. Address issues and resolve conflicts 4. Make sure to reward and recognize members 5. Provide equal opportunities
4. Active listening	1. Stop interrupting 2. Demonstrate empathy 3. Restrain from jumping to conclusions. 4. Seek feedback 5. Follow up on conversations

Leadership Qualities	*Tips for Enhancement*
5. Empathy	1. Showcase genuine 2. Show a willingness to understand 3. Adapt clear communication4. Offer emotional support 5. Continuously seek feed back
6. Emotional intelligence	1. Reflect on emotions 2. Understand the emotions and perspectives 3. Always managing emotions 4. Promote a supportive culture 5. Seek areas for improvement 6. Lead by setting an example
7. Visionary	1. Deeper understanding of organization 2. Strategic foresight to discern long-term opportunities 3. Formulate achievable vision 4. Clearly communicate the vision 5. Inspire and drive the team 6. Regularly adapt the vision based on the feedback 7. Promote a culture of innovation and risk-taking
8. Self-confidence	1. Develop strengths and capabilities 2. Set attainable goals 3. Engage in positive self-talk 4. Undertake new challenges 5. Expand domain of learning and development 6. Surround with a supportive team
9. Ability to motivate	1. Understand the motivations and needs 2. Regularly recognize the accomplishments 3. Provide growth and development opportunities 4. Delegate tasks and empower team 5. Offer constructive feedback
10. Resilience	1. Cultivate a growth-oriented mind-set 2. Maintain a healthy work-life balance 3. Nurture an optimistic outlook 4. Learn from previous experiences 5. Develop stress management and coping strategies 6. Reflect resilience regularly

Leadership Qualities	*Tips for Enhancement*
11. Problem solving skills	1. Encourage garner innovative solutions 2. Encourage group problem-solving sessions 3. Evaluate costs, potential risks, and benefits 4. Empower team members to contribute 5. Undergo regular training and practice
12. Strong communicator	1. Tailor communication styles 2. Communicate with confidence and clarity 3. Encourage a two-way dialogue 4. Demonstrate empathy and emotional intelligence 5. Encourage transparent communication 6. Enhance communication abilities on feedback
13. Strategic mind-set	1. Understand the industry trends and market shifts 2. Promote a culture that values strategic foresight 3. Make decisions based on data analysis 4. Clearly communicate the strategic vision 5. Review regularly 6. Invest in self-professional
14. Leads by example	1. Align actions with vision, values, and goals 2. Demonstrate a strong work ethic 3. Treat team with respect, empathy and fairness 4. Adhere to ethical and professional standards 5. Celebrate successes that contribute 6. Seek feedback and use it
15. Adaptability	1. Harbour a growth-oriented mind-set 2. Be calm and composed when faced with challenges 3. Celebrate small wins and learn from setbacks 4. Communicate transparently about changes 5. Lead by example, demonstration

1.6 Capacity Building based on Structure

The three main categories of training based on structure are as follows.

A. Structured
B. Semi-structured
C. Unstructured

A. Structured Training

The structured training programme is most commonly used by the formal education system. In the structured training programme, training programme is designed by the training institution, considering its relevance to the type of trainees. The programmes are properly structured and covered within the stipulated time period. These training programmes are planned well in advance by convening an expert group meeting held by the training coordinator of an institution.

B. Semi-Structured Training

The semi-structured training programmes are formulated by keeping in view that the training curriculum may be changed to suit the needs of the trainees during the course of a training programme. In such a training programme.

C. Unstructured Training

In unstructured training programmes, trainees themselves structure the training programme schedule with the trainer. In this type of training programme, the trainer has to work hard and plays a critical role for the effective implementation of the training programme. Here, the objective of the programme must be clear both to the trainers as well as to the trainees.

1.7 Capacity Building Based on Context

Training is contextualized when it is based on the experience of the clientele who needs training. Broadly, it comprises the following types.

A. Orientation Training

Provides general information about the organizational structure, service rules, code of conduct, various facilities provided by the organization, *etc.* After going through this training, the employees became fully aware of the whole system of the organization.

B. Induction Training

This training is about the work that is to be performed by the employee. It is more professional and completely job oriented as compared to orientation training. It instructs the employees their duties and responsibilities, and shows how to perform them. Its objective is to prepare the employee to perform job effectively and efficiently.

C. Refresher Training

These trainings are conducted from time to time to keep the employee continuously updated in their professional knowledge and skills. Since scientific advancements and developments are very fast, it is essential for successful organizations to keep their employee informed and trained in such developments.

D. Other Types of Training

(a) **Direct contact training:** In direct contact training programs, trainers and learners work one on one. This limits the number of participants in a program. The direct contact trainings are on campus training or classroom training.

(b) **Distance training:** This training has the advantage of letting workers learn at their own pace, and greatly enhances the ability of different organizations to collaborate in the development and delivery of training programmes. The rapid spread of electronic technologies such as video, closed circuit and responsive television, and rapid data processing for virtually instantaneous feedback has made distance learning highly attractive. Multimedia methods are, now days, used for the successful conduction of distance training.

(*Later steps include: preparing a series of modules to be used by an individual or small groups of participants as progressively advanced learning experiences; preparing ways of reviewing the progress of participants and giving feedback to them on what they have been able to learn well, and where they need to improve further; supplementing cognitive learning with other methods; and reviewing the final learning outcome.*)

(c) **Formal training:** This training takes place in a classroom situation with strict structured time and curriculum schedule. Formal training system was traditionally the most visible and recognised system in the labour market.

(d) **Non-formal training:** In this training the emphasis is on conscious, guided experiences. Many innovative devices are used to convert routine experiences into learning opportunities and also to create new experiences from which participants can learn.

(e) **Centralized organizational of training**: Training modalities can also be differentiated along organizational lines. In a simple centralized model, for instance, one institute is formally charged with planning programs, staffing, and funding them, ensuring their conformity to standards and, often, selecting participants. This pattern still predominates in developing countries with just sufficient training resources for one or two central institutes. The central institute in this design initiates, guides, and supports all aspects of training, and the other units, subsidiary to it, carry out its plans.

(f) **Dispersed organizational training:** Such trainings are conducted at different locations of the main organization taking distance, local needs and resources persons in view.

Types of Training in Short

Structure	*Capacity building*	*Other types*
1. Structured	1. Orientation	1. Direct
2. Unstructured	2. Induction	2. Distance
3. Semi structured	3. Refresher	3. Formal
		4. Non-Formal
		5. Centralized
		6. Dispersed

Training - An Overview

At the other extreme are networks of largely autonomous training units, each responsible for its own training plans and programs and for linking with other units as needed or desired. What and how participants learn in a chosen field then depends, also, on the dispersed model, on which unit they have joined. In developing countries mixed models are increasingly becoming common, in which a central institute sets frames of training, such as priority for training objectives, general guidelines for methodologies and materials to be used, and quantitative parameters for participants, time, and costs. Subsidiary units do the detailed planning, staffing, and management, and contribute actively to formulating and revising system-wide frameworks.

Essential Requirements of Training

Whatever may the type of training, the following are the essential requirements in training instructional plan.

a. Title
b. Trainees Requirements
c. About Training/Background
d. Instructional Material and Training Aids
e. Learning Outcomes
f. Logistics
g. Dates and Duration
h. Contents – Topics, methods, trainers *etc.*
i. Target Learners
j. Feedback

Components of Training

The important components of training are:

i. Goal
ii. Motivation
iii. Opportunity to learn
iv. Opportunity to practice
v. Follow up
vi. Re-training

An Overview Goal

A training programme must have a goal. The goal can further be subdivided into objectives. The objectives of the training will indicate what types of subjects and training methods and technologies to be used in the training programme. Motivation: A participant's motivation is a key element in a successful training programme. Trainees must be willing to invest time and energy to develop them. Motivated employees actively pursue learning to improve their skills and competencies for improving their overall performance.

Training Cycle

Training cycle is an important aspect of training. It is defined as a series of steps or stages that comprise a complete training programme. Training cycle consists of five phases which are narrated below.

1. Training Needs Assessment (TNA)
2. Training Plan Development (TPD)
3. Design of Training Programme (DTP)
4. Training Programme Implementation (TPI)
5. Training Monitoring and Evaluation (TME)

Phase I - Training Needs Assessment (TNA)

In the first phase, the Training Needs Assessment (TNA) phase, the training needs are assessed at six levels of aggregation, namely:

1. Individual
2. Group
3. Organizational
4. State
5. Regional
6. National

A number of methods for training needs assessment are available for different levels of aggregation. Two critical parameters usually in the training needs assessment (TNA) exercise. The first is the national priorities, laid down by policy-makers and planners. The training needs have to be assessed within the ambit of national priorities and programmes. The second is trainees felt needs for training. It is essential to realize the trainees felt needs for training.

Phase II - Training Plan Development (TPD)

After assessment of training needs (Phase I), appropriate training plans are required to be developed in the second phase. A training plan is a manpower document developed on the basis of assessed training needs. It takes into account the existing manpower stock, projects manpower needs for future and quantifies the training requirements. On the basis of assessed training needs for each category of personnel to be trained, it indicates the dates, duration, subject and the institution for training.

A training plan can be (a) Short term (b) Long-term (c) both.

An Annual Training Plan is an essential requirement of sound training management, it is desirable that training plans are developed in, say, a 5 year framework. The singular advantage of developing training plans in a 5-year frameworks is that management will have a reduced task of undertaking an annual review only, enabling it to devote more time to other aspects of training management, which will enhance the value of training.

A Training Plan can be:

a. Territorial - National/State/District
b. Sect oral - Agriculture, health, education
c. Project-related - a comprehensive agricultural development project, which cuts across several sectors
d. Special plan - Training of new recruits to development work
e. Combination of above

Components of Capacity Development

Key components often include developing plans for staff development, driving process improvements, fostering partnerships, and upgrading technology. By implementing a capacity building plan, organizations can better respond to capacity challenges, seize opportunities, and ensure long-term growth and success.

Capacity Pillar

1. Structure
2. Competencies
3. Management systems
4. Enabling policies
5. Knowledge and learning and lastly
6. Leadership.

Ten Key Components of Capacity Development

1. Strategic Leadership Development
2. Technical Proficiency Enhancement
3. Adoption of Smart Technologies
4. Continuous Learning Culture
5. Effective Communication Skills
6. Resource Management Optimization
7. Risk Management Proficiency
8. Customer Relationship Management
9. Sustainability Integration
10. Adaptability to Regulatory Changes

Capacity Development Cycle

The five steps of UNDP's capacity development process are: 1) engage stakeholders on capacity development; 2) assess capacity assets and needs; 3) formulate a capacity development response; 4) implement a capacity development response; and 5) evaluate capacity development.

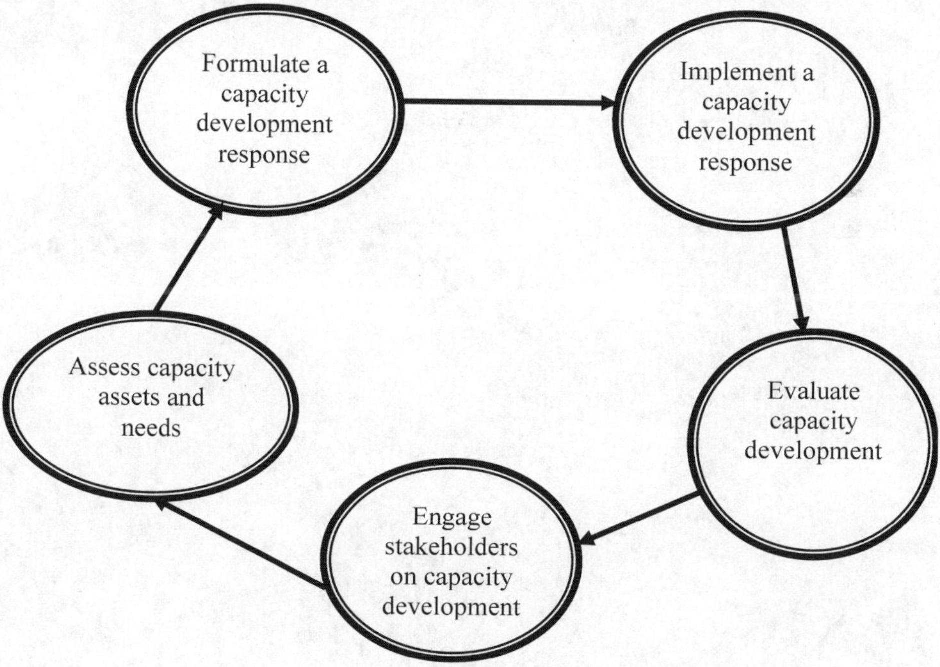

Capacity Development Cycle

Chapter 2

Capacity Development

This Chapter Contains

Sl.No.	*Contents*
2.1	Difference between Approaches and Strategies
2.2	Dilemma- Theory versus Practice, Trainee versus Task, Structured versus Unstructured, Generic and Specific
2.3	Approaches in Capacity: Approaches in Capacity Development -Informative Approach, Participatory Approach, Experimental Approach, Experiential Approach, Performance based Approach
2.4	Capacity Development Strategies: Academic Strategy, Laboratory Strategy, Activity Strategy, Action Strategy, Personal Development Strategy, Organizational Development Strategy

2.1 Difference between Approaches and Strategies

While an approach is informed by ones' view or perspective on issues, a strategy is a carefully planned activities use to achieve a goal; a method becomes the procedure one adopts in solving an issues whiles a technique becomes ones' own unique ways of solving a situation at hand.

The terms "strategy," "approach," and "technique" are often used in the context of planning and execution, especially in fields such as business, project management, and problem-solving. While these terms are related and can sometimes overlap, they have distinct meanings:

Strategies

Are typically broad in scope, long-term in nature, and focused on achieving overall objectives. They provide a framework for decision-making and guide the allocation of resources to achieve a competitive advantage or desired outcome.

Approach

An approach is a more detailed plan or method used to implement a strategy or tackle a specific problem. It is a way of doing something or dealing with a situation.

Approaches are more specific than strategies and often involve the steps or procedures that need to be followed to execute a plan successfully. An approach can be thought of as a pathway or route taken to reach a particular goal within the broader framework of a strategy.

Technique

Techniques are specific methods, procedures, or practices used to accomplish a particular task or objective. They are often practical skills or tools that are applied in a systematic way to achieve desired results. Techniques are more granular than approaches and strategies and are often focused on the implementation of specific actions or tasks. They can be thought of as the tools or practices used to carry out an approach effectively.

2.2 Dilemma- Theory versus Practice, Trainee versus Task, Structured versus Unstructured, Generic and Specific

A. Dilemma

A dilemma is a difficult situation in which we have to choose between two or more alternatives. A dilemma offers two possibilities, neither of which is ambiguously acceptable or preferable.

Dilemma "Is the correct spelling, representing a situation of difficult choices, while "dilemna" is a common but incorrect variant. Proper spelling is crucial for effective communication, and understanding.

B. Theory versus Practice

A theory is a proposed explanation of the relationship between two or more concepts, or an explanation for how/why a phenomenon occurs.

Theory is conceptual or theoretical knowledge that aids in thinking through options, while practice is the application of that knowledge to solve practical problems. Practice is the observation of disparate concepts (or a phenomenon) that needs explanation.

Difference between theory and practice are:

1. It is all too easy to explain the concepts of thirst, pain and sorrow in theory, but the person realizes the difference only when he undergoes these experiences in real life.
2. In theory, many assumptions are made to explain the phenomenon and concepts, whereas in real life, there are no assumptions and conditions are always unique.
3. The dichotomy of theory and practice will remain as these two form the backbone of all learning procedures

C. Training versus Task

The training task is designed to exercise cognitive skills whereas an assessment task is designed to measure cognitive skills.

Tasks, in contrast, are activities that have a product, goal, objective or outcome that require using the target language to achieve it, but are not focused on mechanics.

There is a major difference between a training task and an assessment task. The training task is designed to exercise cognitive skills whereas an assessment task is designed to measure cognitive skills.

A major difference between task training and skill training lies in the subject matter presented prior to, during, between or after the practice session(s). For task training, the subject matter is easily confined to the steps of the task, such nomenclature as might be necessary to follow instructions or use a job aid and a demonstration of some kind. The task determines the subject matter. For skill development, we are in a very different situation. We are confronted by the fact that we really don't know what underlies a skill. We have some notions, some theories, some hypotheses, but we really don't know. And so we present subject matter consisting of concepts, models, methods, approaches and techniques to name a few. The practice we provide is practice in applying these artefacts to the situations we set up in the training.

D. Structured versus Unstructured

Structured data is data that fits neatly into data tables and includes discrete data types such as numbers, short text, and dates. Unstructured data doesn't fit neatly into a data table because its size or nature: for example, audio and video files and large text document

Unstructured Data

Unstructured data is more or less all the data that is not structured. Even though unstructured data may have a native, internal structure, it's not structured in a predefined way. There is no data model; the data is stored in its native format.

Semi-Structured Data

Semi structured data is a third category that falls somewhere between the other two. It's a type of structured data that does not fit into the formal structure of a relational data base. But while not matching the description of structured data entirely, it still employs tagging systems and other identifiable markers, separating different elements and enabling search.

Difference between Structured and Unstructured Data

	Structured Data	*Unstructured Data*
Definition	Data is organized and formatted in specific way following predefined model or scheme	Data that lacks specific structure or format typically unorganized or in raw form
Organization	Well organized with specific format such as tables and columns	Lacks predefined format and is unorganized
Accessibility	Highly accessible and can be easily retrieved	Less accessible and requires advanced technology for extraction and analysis
Examples	Customer information, Transaction records, inventory list, financial data	E-mail social media post. Multimedia files
Analysis	Easily analysed using traditional statistical methods and data mining techniques	Requires advanced techniques like natural language processing machine Learning algorithms analysis
Scalability	Limited scalability due to predefined scheme and fixed data structure	Highly scalability and can accommodate any type of data without altering existing structure
Use of cases	Business intelligence data analytic, financial reporting	Sentiment analysis, Social media monitoring, text mining

Five Key Differences

1. ***Structured vs. Unstructured Data***

Defined vs. undefined data: Structured data is clearly defined data in a structure. While unstructured data is usually stored in its native format, structured data lives in rows and columns and can be mapped into predefined fields.

Unlike structured data, which one can organize and access in relational databases, unstructured data does not have a predefined data model and is undefined.

2. ***Qualitative vs. Quantitative Data***

Another difference between structured and unstructured data is that structured data is often quantitative data, meaning it usually consists of hard numbers or things that can be counted.

Unstructured data, on the other hand, is often categorized as qualitative data and cannot be processed and analysed using conventional tools and methods.

3. ***Data Storage in Data Warehouses vs. Data Lakes***

Businesses often store structured data in data warehouses and unstructured data in data lakes. A data warehouse is an end point for the data's journey through an ETL pipeline. A data lake, on the other hand, is a sort of almost limitless repository where we store data in its original format or after undergoing a basic "cleaning" process.

4. ***Ease of Analysis***

One of the most significant differences between structured and unstructured data is how well-structured data lends itself to analysis. Structured data is easy to search, both for data analytics experts and for algorithms. Unstructured data, on the other hand, is intrinsically more difficult to search and requires processing to become understandable.

5. ***Predefined Format vs. Variety of Formats***

The most common format for structured data is text and numbers. Structured data has been defined beforehand in a data model.

Unstructured data, on the other hand, comes in a variety of shapes and sizes. It can consist of everything from audio, video, and imagery to email and sensor data. There is no data model for the unstructured data.

E. Generic and Specific

Generic shared by typical of or relating to a whole group of similar things, rather than to any particular thing. Generic is used to describe something that refers or relates to a whole class of similar things. Parmesan is a generic term used to describe a family of hard Italian cheeses. *Synonyms: collective, general, common, wide.*

A generic product, whether it's a tire or donut or drug, is typical of all other products like it. There's nothing distinctive or unique about it. A generic fan looks and acts just like dozens of other fans, it doesn't spin backwards or send out sparks. Generic aspirin doesn't have a trademark, like Bayer or St. Joseph: it's just plain aspirin.

Specific

Adjective having a special application, bearing, or reference; specifying, explicit, or definite: to state one's specific purpose. Specified, precise, or particular: a specific sum of money.

A specific example is a particular instance. So to give a specific example of technology would be to write about particular people using a particular machine. To give a specific example of any human activity would require that we write about individual people.

2.3 Approaches in Capacity: Approaches in Capacity Development - Informative Approach, Participatory Approach, Experimental Approach, Experiential Approach, Performance based Approach

A. Informative Approach

This approach provides information to the trainees on various subjects. For example, development professionals working in health sector need information about the performance and functioning of various health programmes.

B. Participatory Approach

In the participatory approach, the trainees act as facilitators and, instead of following a one-sided lecturing approach, they involve trainees in the process of teaching and learning. In this process, the trainer also learns from the trainees and teaching and learning becomes a two-way process. The participatory approach empowers the trainees and the trainer, equally. The participatory approach is considered as a democratic way of training and followed by most of the development organizations involved in the training at the grassroots.

C. Experimental Approach

The experimental approach emphasises situations in which the trainees will eventually operate. In this approach, both the trainers as well as the trainees jointly formulate the objectives and methodologies of training programme. Here, the trainers primarily serve as facilitators, catalysts, or resource persons. This method applies more effectively to those who have some field experiences.

D. Experiential Approach

Experiential Learning is the process of learning by doing. By engaging trainees in hands-on experiences and reflection, they are better able to connect theories and knowledge learned in the classroom to real-world situations.

When trainees participate in experiential education opportunities, they gain:

1. A better understanding of course material
2. A broader view of the world and an appreciation of community
3. Insight into their own skills, interests, passions, and values
4. Opportunities to collaborate with diverse organizations and people
5. Positive professional practices and skill sets
6. The gratification of assisting in meeting community needs
7. Self-confidence and leadership skills

E. Performance Based Approach

In the performance based approach, the goal of the training programme is measured by the attainment of a given level of proficiency. The performance based training approach is developed by Elan in 1971. It is mostly task or skill centred, and can be applied to non-formal education and extension education.

2.4 Capacity Development Strategies: Academic Strategy, Laboratory Strategy, Activity Strategy, Action Strategy, Personal Development Strategy, Organizational Development Strategy

Effective capacity building strategies include up skilling and deskilling employees, leadership development, outsourcing resources for flexibility, forming partnerships for added value, and streamlining agency processes for efficiency.

A. Academic Strategy

Academic strategy works on skills surrounding self-knowledge and self-awareness, time management, reading comprehension, short- and long-term memory, self-advocacy skills, critical thinking, test anxiety, amongst others.

The Academic Strategy builds on the aspirations of the College Strategy and department and faculty strategies, and connects up with other strategies including Learning and Teaching, Advancement and Estates. By working together, we can increase our capacity and capability, enabling the kind of reach that can only be attained by working at scale across the College. This will help us to increase our reach and deliver outcomes that both make a difference in the world and raise Imperial's profile.

B. Laboratory Strategy

The laboratory's strategic planning process includes the development of lab goals and objectives, defining lab activities funded for the current fiscal year, and communicating the plan to lab employees, the science community, and funding agencies.

C. Activity Strategy

Strategies are broad concepts or approaches to achieve the project objectives while activities are actions that are undertaken within these strategies. For example, 'building the capacity of the community members' is a strategy, our project has adopted.

Activity based teaching is a method adopted by a trainer to emphasize technique of teaching through action in which the learners take interest comprehensively and realize effective learning practices. It is the procedure in which the trainee is effectively included in taking interest rationally and physically.

D. Strategic Activities

Strategic activities are actions, projects and other activities that are designed to achieve specific outcomes and objectives. Each objective and goal should have a plan (made up of activities) that details how it will be achieved.

E. Personal Development Strategy

These strategies often include setting personal goals, seeking feedback, continuous learning, reflecting on one's experiences, and developing new habits or skills. Leaders, in particular, need personal development strategies for many reasons.

Ten Personal Development Strategies

1. Coaching and Mentoring
2. Empowerment
3. Provide Constructive Feedback
4. Encourage Continuous Learning
5. Set Clear Expectations and Goals
6. Model the Behaviour
7. Foster a Positive Environment
8. Promote Work-Life Balance
9. Recognize and Reward
10. Encourage Networking

F. Organizational Development Strategy

The organizational development strategy also means aligning the goals of the company with an orientation towards people, innovation, knowledge creation, and community-based development.

1. **Planning of organizational development:** This component refers to planning the overall goals and activities for developing the internal and external situation of the company:
    a. Objectives for the organizational development of the company
    b. Capabilities and resources
    d. Budget for the organizational development
    d. Calendar of activities
2. **Key areas of organizational development:** This component refers to mapping the main key areas that form the basis of organizational development:
    a. Employees
    b. Working relations
    c. Work flow
    d. Product quality
    e. Innovation
    f. Community development
3. **Solutions to key areas of organizational development:** This component refers to brainstorming and analyzing of ideas with the teams and the rest of the stakeholders to develop practical and cost-effective solutions for organizational development. The steps are:
    a. Proposed solutions to the key areas of the organizational development
    b. SWOT Analysis of the proposed solutions
4. **Implementation of Solutions:** This component refers to the implementation of the proposed ideas by following a certain schedule.
    a. Schedule of activities
    b. Actual implementation
    c. Monitoring of the implementation
5. **Evaluation of the Results:** This component refers to the evaluation of the overall implementation and results achieved.
    a. Evaluation of the implemented solutions
    b. Evaluation of the overall results
    c. Feedback from the stakeholders involved
6. **Continuous Improvement:** This component represents the maintenance of a dialogue and knowledge sharing atmosphere inside the company to improve the process of organizational development.
    a. Dialogue with the stakeholders
    b. Informal debates
    c. Knowledge sharing

Chapter 3

Planning and Organization of Capacity

This Chapter Contains

Sl.No.	*Contents*
3.1	Steps in Designing and Planning of Capacity Development
3.2	Organisation of Training
3.3	Stakeholders in Training and their Responsibilities

3.1 Steps in Designing and Planning of Capacity Development

The capacity planning process is characterized by:

1. To determine production
2. Planning to meet changing needs for the product
3. Ability of organization to complete work within stipulated time
4. Refers to ability of organization to accomplish goal
5. Balancing available resources to meet satisfaction of the customers.

For above reasons we opt for designing and planning of capacity development. In other words we design training plan.

A training design is a blueprint of a training event experience. It is a detailed plan of what we will do, why we will be doing it, and the best way to reach our training objectives. The following is an eight step model for designing an effective training event that meets the needs of participants and trainers alike.

Step 1: Select the participants of the training and target audience
Step 2: Determine the participants' needs
Step 3: Formulate training goal and objectives
Step 4: Outline the training content.
Step 5: Develop instructional activities
Step 6: Prepare the training design
Step 7: Prepare a participant's evaluation form
Step 8: Determine follow-up activities

Step 1 - Select the participants of the training and target audience

Before selecting the participants for training, the organisers of the training programme have to introspect about some fundamental questions. The basic considerations to select participants for training are, specific benefit, need of the participants to attend training and for what specific purpose the organization would depute a person for training.

Step 2 - Determine the participants' needs

The specific training needs of the participants will influence the development of learning objectives and guide the choice of activities and training strategies. The more we know about participants, the greater will be training management and effectiveness. Accordingly we can design the training events.

Step 3 - Formulate training goal and objectives

After assessing the needs and expectations of the participants, we are ready to define the goals and objectives for the training. Objectives should be more precise, specifying a path for achieving the programme goals. We can follow the following steps to formulate training goal.

i. List the ways we would like training to benefit the participants
ii. Write down those outcomes as objectives.
iii. Check objectivities from the perspectives of the training participants.
iv. Set training priorities. Rank objectives according to their importance.

Step 4 - Outline the training content

Most training divided into three key segments: introduction, learning core component and the wrap up and evaluation segment.

(i) Introduction: The introduction establishes a positive learning environment. The opening activities should stimulate interest and enthusiasm in the training and reduce anxiety among participants. It is important to build some content in the introduction, so that participants experience these activities as meaningful.

(ii) Learning core component: This is the body of the programme, participants engage in activities designed to accomplish the training objectivities; concepts and ideas are taught and explained; resources are shared; and, training strategies and skills are demonstrated, practiced and discussed. To be most effective, activities should actively involve participants in acquiring knowledge or practicing skills.

(iii) Wrap up and evaluation segment: This segment should help bridge the gap between training and training implementation and promote a positive feeling of closure.

Step 5 - Develop instructional activities

The development of training design consists of organizing learning activities, so that outcomes identified by the objectivities are achieved. Make sure that activities flow logically from one to the next. As we develop activities, select training strategies that are most likely to help to meet objectivities. For example, skill development is best achieved through modelling, practice, and feedback, while information acquisition can be achieved through group discussion.

Step 6 - Prepare the training design

Create a written document that provides a detailed plan of the training sessions, including goals and objectives; the sequence of specific learning activities and the time allotted to cover during each activity; and, consider the trainer who will be responsible for the activity take into account the skill, expertise, training style, comfort level with the subject, *etc.*

Step 7 - Prepare a participant's evaluation form

The purpose of evaluation is to determine the extent to which the training achieved its objectives and to identify what adjustment, if any, needs to be made to the training design or to follow–up process.

Step 8 - Determine follow-up activities

Without follow–up, the benefits of training may quickly be forgotten or never used. Follow up activity provides the continued support and feedback necessary for successful implementation of new ideas and practices. To be effective, follow– up should be planned as we develop our training design, and should include a range of opportunities for participants to reflect on both the content that they learned during the training, and the process of implementation.

3.2 Organisation of Training

The training organisers have to take care of many operational arrangements at different stages of organisations of training programme *i.e.*

Organization of Training

1. Before the training programme
2. During the training programme
3. Middle of the training programme
4. At the end of the training programme
5. After the training programme
6. Follow up

1. Before the Training Programme

The training organisers have to carefully look after following aspects before the training programme.

i) Check the materials.
ii) Receiving the trainees.
iii) Lodging of trainees.
iv) Arrangement for registration
v) Boarding arrangements
vi) Meeting of Trainers/Faculty

2. During the Training Programme

The following steps needed to be taken during the conduction of the training programme.

i) Inauguration of training programme
ii) Pre-test: The pre-testing of existing knowledge level

iii) Training need assessment

iv) Ice-breaking sessions

3. Middle of the Training Programme

During the middle of the training programme the following actions need to be taken.

i) Power failure

ii) Guest speakers

iii) Equipment for training

iv) Prompt action

v) Timely information

vi) Arrangement for lunch

vii) Preparation of lecture note

viii) Identification of field applicable technology

ix) Provision of additional information, if required

ix) Cordial relationship between trainees and trainers

x) Hand-outs to trainees

xi) Monitoring of the course activities

xii) Informal feedback from the trainee

xiii) Observation about feeling of the trainees

xiv) Review session

xv) Course certificate

xvii) Group photograph

4. At the End of the Training Programme

At the end of the programme course director has to take care of following aspects.

i) Back home application

ii) Post-test for knowledge gain

iii) Training evaluation

iv) Concluding session

v) Return journey

5. After the Training Programme

Training is a continuous process which ends when an objective has been achieved.

i) Preparation of training programme evaluation report

ii) Follow-up program
iii) Settlement of accounts
iv) Follow-up program

6. Follow-Up

Following is one of the important but utterly neglected aspects of training. The follow–up of a training programme can be defined as the activities that identify the utility of the programme in job performance; the reasons for non – utilization of new skills or knowledge acquired during the training, and; providing help and reinforcement for the application of the new learning, as well as an assessment of future training requirements.

Need for Follow-up	*Benefits*
1. Importance of follow-up	i) Identifies the actual impact on individual job ii) The feedback about the utility of iii) For actual realization of impact of raining iv) To get feedback from the participants about training.
2. Objectives of follow up of a programme	i) Extent of utilization of newly acquired knowledge ii) Reasons for poor transfer of learning if any. iii) Up keeping motivation level iv) Feedback on impact of the training v) Assessment the future training needs vi) Sharing of experiences vii) Timing of follow up
3. Components of a follow–up system	i) An analysis of utility of the programme ii) Identification of the reasons for non- adoption of newly acquired skills and knowledge.

Reinforcement of the motivation and skills of the participants through;

i) Post-training contact through a regular newsletter, journal *etc.*
ii) Refresher courses
iii) Personal contact
iv) Support for the adoption of newly acquired skills is provided through
v) Clarification of doubts,
vi) Making arrangements for the availability of equipment's, materials
vii) Providing constructive feedback
viii) Encouraging interaction among the trainees through

ix) Formation of alumni association

x) Development of e-mailing list, for network among the trainees

3.3 Stakeholders in Training and their Responsibilities

In any training programme, along with the participants and the training team, the training agency and the participating organizations are among the key stakeholders. If the funding for the programme has been provided by a donor agency, it too becomes a stakeholder, depending on the nature of its involvement in the training. The followings provide an overview of key functions or responsibilities of these stakeholders

1. Participants

i) Participate and contribute to all training activities. Set out personal learning objectives and direct efforts to realize them

ii) Extend supports to the training team in achieving the programme objectives and maintain a learning environment.

iii) Assist co-participants in realizing their learning objectives

2. Organizations Work

i) Assist the training agency in identifying training needs of their nominees.

ii) Select appropriate participants in accordance with the criteria and requirements set out by the training agency.

iii) Communicate expectations from the programme to their nominees and the training agency.

iv) Keep in regular contact with the training agency to secure information on the progress and performance of their nominees, especially if it is a long term programme

v) Provide appropriate opportunities and support to their participants to facilitate the transfer of training to the field situation.

vi) Maintain the performance or actions of their nominees after the training programme and provide necessary data to the training agency

3. Training Institution

i) Provides an environment that is conducive to learning, including physical facilities.

ii) Ensures effective coordination between the trainers and the support staff associated with the delivery of the programme.

iii) Maintains contacts with the sponsoring agencies, especially in a long term programme, and keeps it informed of the progress and performance of its nominees.

iv) Maintains consistency between what it advocates in the programme and what it practices in its own functioning. 46 Fundamentals of Training

v) Assists in the professional growth and development of the training staff.

vi) Maintains the training team; keeps pace with the developments in the field of training

vii) Appraisal of work on a regular basis.

Chapter 4

Capacity Development Programmes

This Chapter Contains

Sl.No.	*Content*
4.1	Concept of Need Assessment
4.2	Approaches in Need Analysis- Performance Analysis, Task Analysis, Competency Study and Need Survey

4.1 Concept of Need Assessment

The assessment is designed to assess individual capacity gaps so training and development plans can then be prepared to address them. Increasingly, the dimensions of accountability and incentives, performance, values and ethics are becoming even more important at the level of individual capacity assessments.

Defining Capacity Development (UNDP 1997)

An essential feature of capacity assessment is a common understanding of what capacity and capacity development means. This is not as straightforward as it sounds. To some it is synonymous with workshops and training, to senior managers it can mean organizational development, to non-governmental organizations (NGOs). It is associated with empowering individuals and grassroots organizations and to international agencies and donors.

Clarifying the meaning of capacity development was one of the main issues addressed at the ICID/FAO workshop in Montpellier (FAO 2004). This workshop followed the work of UNDP and defined capacity development as' '*the sum of efforts needed to nurture, enhance and utilize the skills and capabilities of people and institutions at all levels - locally, nationally, regionally and internationally - so that they can better progress towards sustainable development at the basic conceptual level, building capacity involves empowering people and organizations to solve their problems, rather than attempting to solve problems directly. When capacity development is successful, the result is more effective people and institutions better able to provide products and services on a on the sustainable basis.*'

On the basis of above narrations there are three levels of capacity development.

Level I: At national and international level referred as the enabling environment.

Level II: At the organizational level, which refers to the wide range of organizations involved in different projects.

Level III: At the individual level, which is the most structured and familiar part of capacity development and includes education and training of the various stakeholders, from farmers to local professionals.

The three levels provide a structure that allows capacity development to be examined and analysed.

The Process of Capacity Development

There are five strategic phases of capacity development.

Phases	*Narration*
First	What is present or existing capacity in the system? Through base line survey it can be established. The question is asked "Where do we stand now? or, Where we are now?

Phases	*Narration*
Second	What should be future desired capacity development leading to question where do we want to go?
Third	Gap in between two *i.e.*, existing level and desired level of capacity development to decide strategic plan to indicate," How do we get there?"
Fourth	Implementation phase to meet objectives, "What actions do we take?"
Fifth	Evaluation of result, what is achieved and what is missing

Of the five phases, assessment of capacity needs, that is, establishing the existing and required capacities as well as identifying the gaps between both, is perhaps the least well developed, at the same time it is the most vital. Without a proper understanding of what currently exists and what is needed, there is a good chance that inappropriate measures and actions will be initiated.

The UNDP guidelines introduce a note of realism by suggesting that putting this concept into practice is not a simple process. The first step of assessing existing capacity can usually produce much useful information. The second step of looking at future capacity, is less sure as it largely depends on policies and strategies for future development and these may not always be as clear as they should be. This has a knock-on effect on the third step of assessing the gaps and hence on developing strategies to fill the gaps.

UNDP makes the case for capacity assessment as a structured and analytical process, whereby the various dimensions of capacity are assessed within the broader socio-economic environment, as well as evaluated for specific organizations and individuals. They indicate how assessments might be undertaken in different situations and how programmes and projects might be better designed to ensure ownership, sustainability and ultimately success. Special emphasis is given to using existing capacities as the basis for capacity development.

Enabling Environment

Assessments at this level include the dimensions of policy, legal and regulatory framework, management and accountability, resources and processes (the inter-relationships between organizations) and how resources and information flow both formally and informally. For development initiatives that are national in context (*e.g.* governance or environmental programmes), all the dimensions would need to be examined. However, for initiatives at a sectorial level, the analysis would include only the relevant components.

Matrix for Assessing Capacity (UNDP 1997)

Sl.No.	*Dimensions of Capacity*	*Existing Capacity*	*Possible Future Capacity*	*Estimated Capacity Gap*	*Possible Strategies*
1.	Environment - Level				
2.	A Policy framework				
3.	Legal and regulatory framework				
4.	Management accountability				
5.	Resources				
6.	Processes and relationships				
7.	Dimensions of capacity Organization - Level II				
8.	Strategic management				
9.	Culture/structure				
10.	Processes				
11.	Human resources				
12.	Resources - financial				
13.	Resources - information				
14.	Infrastructure				
15.	Dimensions of capacity Individual - Level III				
16.	Job skills and needs				
17.	Professional development				
18.	Access to information				
19.	Performance/incentives				
20.	Values/attitudes/motivation				
21.	Relationships/				
22.	Professional integrity				

Mapping Capacity and Finding the Gaps

The rows represent the various dimensions of capacity. In most situations, an assessment will generate information on existing capacity. Moving to the next step of assessing future capacity may be less certain, information on this will only emerge gradually as answers to the question - *where do we want to be?* - emerge.

Existing Capacity

A critical issue for undertaking an assessment is the entry point. Once an entry is made, movement can then take place from one level, enabling environment, organization or individual, to another according to needs. Some information, gathered from the assessment at a certain level, might indicate that another level should be assessed.

Future Capacity

Assessing future capacity can be a difficult and time consuming step, as stakeholders may not always have a clear vision of what they would like to achieve. A realistic assessment of development objectives, resources and timing is required. A general consensus among all the stakeholders involved is also needed if the goals are to be achieved.

Tools for Assessment - Establishing the Baseline

A range of tools and approaches exist for assessing capacity at the different levels.

The Enabling Environment

At the level of the enabling environment, the two fundamental needs are to map and understand the policy environment and to assess the legal and regulatory framework. Once these are understood, the other issues to be considered at this level, such as accountability, resources, processes and relationships, can usually be identified and mapped with relative ease.

1. **Policy characteristics analysis**: It assists in identifying the driving forces, decision-makers, costs and benefits, and those who bear the costs and benefits of policy-making in particular sectors.
2. **Political mapping**: It identifies stakeholders in policy-making, differentiating between external stakeholders, social groups, political parties and pressure groups, and distinguishing their different levels of support and influence, and
3. **Policy network maps**: The policy try to graphically present the linkages between different aspects and sectors in policy-making, in order to identify inconsistencies and constraints and to support so-called 'joined-up government'.

The Organizational Level

Assessment of existing capacity at the organizational level must take account of these organizations within their institutional context and consider their linkages and relationships with other relevant organizations. A first step in capacity assessment at the organizational level is therefore some form of stakeholder analysis. There are three approaches,

1. Open system model, which considers the strategic and institutional environment, the organizations inputs and resources, its strategy, culture, human resource management policies, systems, structure, outputs and performance;
2. The 7-S model - strategy, structure, systems, shared values, skills, style and staff; and
3. Organizational elements model covering inputs, processes, products, outputs, and outcomes.

It can be seen that all cover approximately the same ground, including most of the components itemized in the sublevels of the assessment matrix. All can be presented in the form of a table or matrix, making for easier visual understanding.

Individual Level

Techniques for assessing capacity at the individual level are now well established. Indeed much of the criticism of capacity development programmes in the past is that they have been focused solely at the individual level, without paying sufficient attention to the organizational framework within which the individual operates. Individual capacities are often assessed in relation to the need for training and lead to some form of assessment of training needs.

General Assessment Tools

Certain assessment tools can be used at all three levels of capacity assessment. These include:

1. **Problem tree analysis:** It is used to identify causes and effects of identified problems, and helps to look beyond visible symptoms to reach an understanding of underlying causes. The problem tree analysis may also be turned into an objective tree analysis, which forms the basis of a strategy to deal with the identified problems.
2. **SWOT analysis:** Strengths, weaknesses, opportunities, threats (SWOT) analysis can be used by stakeholders to assess the enabling environment, the institutional framework and the specific organization; and
3. **Benchmarking:** This, method is increasingly used in the irrigation and drainage sector as a way of comparing performance and output over time, or with other organizations. It is a specific and structured method of performance assessment, which in its more general form can be used as a basis of assessing existing capacity without making comparisons with others or over time.

4.2 Approaches in Need Analysis: Performance Analysis, Task Analysis, Competency Study and Need Survey

A. Performance Analysis

Definition

A Performance Need Analysis is the process of identifying the gap between desired performance and actual performance. A Performance Need Analysis is a much broader analysis as it identifies several possible performance issues including lack of motivation, lack of resources, lack of knowledge, or ineffective processes.

A performance needs analysis (PNA) is a process of identifying and addressing the gaps between the current and desired performance of learners. It helps to design effective and relevant learning solutions that align with organizational goals and learner expectation. It is conducted in six phases.

1. **Define the problem:** The first step of a PNA is to define the problem that we want to solve with learning solution. We need to clearly articulate the performance issue, its impact, and its root causes. We can use various methods to collect data, such as surveys, interviews, observations, or document reviews.
2. **Determine the desired outcome:** The next step of a PNA is to determine the desired outcome that we want to achieve with learning solution. We need to define the performance indicators, the criteria, and the standards that will measure the success of our intervention. We should also align outcome with organizational objectives and learner needs. We can use tools such as SMART goals, logic models, or action plans to specify outcome.
3. **Analyse the performance gap:** The third step of a PNA is to analyze the performance gap that exists between the current and desired performance of learners. We need to quantify and qualify the gap, and identify the factors that contribute to it. We can use frameworks such as the Kirkpatrick model, the ADDIE model, or the SWOT analysis to analyze the gap. We should also prioritize the gaps based on their urgency, importance, and feasibility.
4. **Identify the possible solutions:** The fourth step of a PNA is to identify the possible solutions that can address the performance gap and achieve the desired outcome. We need to consider the different types of interventions, such as training, coaching, mentoring, feedback, or incentives, and evaluate their pros and cons. We should also consider the resources, constraints, and risks that may affect solution. We can use tools such as brainstorming, decision matrices, or cost-benefit analysis to identify the best solution.

5. **Design the learning solution:** The fifth step of a PNA is to design the learning solution that we have selected as the most suitable intervention. We need to follow the principles of instructional design, such as analysis, design, development, implementation, and evaluation, to create an effective and engaging learning experience. We should also apply the best practices of adult learning, such as relevance, motivation, feedback, and reinforcement, to ensure solution meets the learner needs.
6. **Evaluate the results:** The final step of a PNA is to evaluate the results of learning solution and its impact on the performance gap and the desired outcome. We need to collect and analyse data, such as learner satisfaction, knowledge retention, skill transfer, or business results, to measure the effectiveness and efficiency of solution. We should also solicit feedback from the stakeholders, such as learners, managers, or customers, to identify the strengths and weaknesses of solution and make improvements.

B. Task Analysis

Definition

Task-analysis steps are completed to determine which aspects of the job to include in a training program: Identify tasks that are required for performing the job. For each task, list the exact steps that a person must do to complete the task. List every step from start to finish in order, no matter how small the step

At some point in our career we're likely to find folks who are involved in task analysis. In its simplest form, task analysis breaks a job (task) down into observable steps. It also reveals the knowledge and skills the employee needs to complete each step.

At times task analysis may be used to determine training needs. It may be a part of the analysis completed during the needs assessment. Learning objectives may be written based on the task analysis.

How to Conduct Task Analysis?

1. Identify tasks that are required for performing the job.
2. For each task, list the exact steps that a person must do to complete the task. List every step from start to finish in order, no matter how small the step.
3. When listing the steps, make them as clear as possible. Even the smallest assumption can create confusion in the actual performance of the task.
4. After we have completed all steps for every task, go back and identify each step in one of three ways:
   (i) **Common knowledge:** Trainee will know how to do this step because of common knowledge. No training is necessary for this step.

(ii) **On-the-job training:** The step is simple enough that the trainee can learn it on the job. No training is necessary for this step, but we may want to consider developing a job aid such as a check sheet or a list of procedures.

(iii) **Training topic:** The trainee will not know how to do this step without training. The technical training program should be based around these topics.

(iv) **Ask a subject matter expert (SME)** to review task analysis to see whether we have analyzed the job(s) correctly.

(v) **Design training program around those topics we identified as requiring training:** Task analysis and training objectives are used prior to designing the training program. Both are critical steps between conducting the needs assessment and designing the training program. Both play a part in setting the goals of the training session and ultimately achieving them. As David Campbell says, "Aim at nothing and we'll hit it every time".

C. Competency Study

A training needs assessment identifies individuals' current level of competency, skill or knowledge in one or more areas and compares that competency level to the required competency standard established for their positions or other positions within the organization.

What are the three levels of training needs analysis? There are three sources used to analyse training needs: organizational analysis, operational analysis, and individual analysis.

Competency Mapping

Competency mapping is an effective methodology to incorporate the competencies throughout the various processes and to assess the training needs of the organization. It is the most systematic and scientific method of training needs assessment, wherein, the starting point is to understand the job requirements. Thereafter, measuring the competency levels of the employees, the gaps can be identified between the competencies desired and current state of competencies.

Meaning of Competency

Mapping David McClelland (1973) proposed the concept of competencies that are required for effective job performance. He argued that being knowledgeable and/or intelligent only does not indicate that a person is an effective and efficient worker- a worker's performance is a function of his/her knowledge plus skills and attitudes. One should not be judged solely on how knowledgeable they are in their technical subject area of expertise but on how skilful and able they are in delivering services to their clients.

Skill is the ability to perform the job and competency is the requirements of skill set for that particular job. In other words, skills are specific activities, and competence is the ability to carry out an activity effectively, safely, and efficiently. The most critical competencies are those that relate to skills that a development worker is expected to perform. It should also be noted that core competency needs are contextual, and development workers' contexts affect their competency needs and competency levels.

Steps in Competency Mapping

Sulthana and Madhavaiah (2017) suggested three steps in Competency Mapping:

First: A job analysis is carried out by asking employees to fill in a questionnaire that asks them to describe what they are doing, and what skills, attitudes and abilities they need to have to perform it well.

Second: Having discovered the similarities in the questionnaires, a Competency-based job description is crafted.

Third: Having agreed on the job requirements and the skills and attitudes needed to progress within it and become more productive; one starts mapping the capability of the employees to the benchmarks. It helps in determining the training and development needs.

Methods of Competency Mapping

Common competency mapping methods/approaches suggested by Shari Parsons Miller (2018) are

1. Assessment centre
2. Critical incidents technique
3. Interviewing
4. Questionnaires and
5. Psychometric tests.

Assessment Centre

Let us discuss them briefly hereunder (Shari Parsons Miller, 2018). Assessment Centre In the assessment centre methodology the demands of the job, performance and training needs are mapped in relation to specific job attributes by situational observations, various types of discussions and simulation exercises. The employees are asked to work through certain situations and their behaviour is observed. Finally the current and future requirements of the job needs are assessed by mapping them with skills and aptitude of the employees.

Critical Incidents Technique

In critical incident technique, the data about critical events *viz.*, very good performance or complete failures are collected along with possible reasons through interviews or questionnaires as soon as the job is performed. The incidents are then categorized according to associated job behaviours to reveal patterns of performance gaps, strengths and finally mapped with the training needs.

Personal Interviews

The behaviour of employees is judged with targeted interview questions related to concrete job experiences, general motivation, disposition and outlook. The responses of the interview are compared against the competency map created for the job to determine which employee offers the best match for the organization's needs

Questionnaires Job Competencies

The work performance are assessed by questionnaire method. The data collected through questionnaires includes:

1. Work requirements and activities
2. Functional job analysis that describes job duties and characteristics
3. Occupational analysis inventory that specifies work elements for almost all occupations
4. Position analysis that ties job characteristics to human characteristics, and
5. Work profiling system that measures ability and personality attributes. The multipurpose occupational system analysis inventory collected through questionnaires highlights tasks and competencies for jobs.

Psychometric Tests

The behavioural attributes can be quantified through psychometric testing. The most commonly used psychometric assessment tests are:

1. **Aptitude tests:** Aptitude tests identify natural liking in a specific job area, and are designed to help predict how well a person would perform in a given job after being provided with training
2. **Achievement tests:** Achievement tests measure the level of proficiency achieved in a certain job area after training.

Task analysis is a systematic analysis of jobs to identify job contents, knowledge, skills and aptitudes required to perform the job. It is an effective and widely used approach for determining training needs in which training needs are linked to task performance.

Usefulness of Task Analysis

Helps to

1. Determine the learning goals and objectives;

2. Define and describe in detail the tasks and sub-tasks that the employee will perform;
3. Specify the knowledge type (declarative, structural, and procedural knowledge) that characterise a job or task;
4. Select learning outcomes that are appropriate for training;
5. Prioritise and sequence tasks;
6. Determine learning activities and strategies that foster the required development;
7. Select appropriate media and learning environments;
8. Construct performance assessments and evaluation

Task Identification or job analysis refers to dividing a major work event into its component parts, which allows a trainer to understand what an individual does in the course of his/her work. Associated activities required to be accomplished for task identification are narrated below.

Assumptions in Task Identification

1. A problem is nothing but a gap between desired performance and actual performance.
2. Not all discrepancies between desired and actual behaviour are remedied by designing and delivering a training programme. For example, the problem could be due to poor organizational policies or environment.
3. The need identification process helps the trainer to distinguish training and non-training problems with their causes, and then, to design training, matching to training problems.

Methods in Task Identification

Commonly used methods for task identification are:

1. Interview
2. Questionnaire
3. Case study, and
4. Observation method

Steps in Task Identification

Step 1 Identify and define the job or major work event for analysis. Job analysis may not always involve an analysis of the entire job, but, often a limited area of a job.

Step 2 Prepare a list of all the tasks involved in it using any of the following methods:

1. Seeking expert help

2. Use of techniques like observation and interview
3. Meeting and group discussion
4. Case method

Step 3 Verify the identified tasks with the help of experts, supervisors and a selected group of employees.

Step 4 Make the final list of tasks.

Method Used to Priority the Task

There are three dimensions to priority the task need identification. These are, performance frequency, relative importance of the task and degree of difficulty to learn the kill for betterment. Examples are given with score analysis methods to ascertain priority of task training needs.

Frequency of Performance		*Relative Importance of Task*		*Degree of Difficulty to Learn*	
Frequency	*Score*	*Frequency*	*Score*	*Frequency*	*Score*
Rarely	1	Slightly important	1	Easy	1
Occasionally	2	Moderately important	2	Difficult	2
Fortnightly	3	Highly important	3	Very difficult	3
Weekly	4				
Daily	5				

Work out the total score for each task by adding the scores for frequency, relative importance, and extent of difficulty in learning.

The priority areas for training can be identified based upon the total score for each task.

Skill Gap Analysis

According to Halim and Ali, the skill-gap analysis is a process of determining the training needs of individual employees in relation to the important tasks steps or components of tasks identified for training. It determines how skilled or proficient individual employees are on these task-steps or components, how many individuals differ from desired performance and whether or not they need training.

The steps used in skill gap analysis are as follows.

i) Prepare a priority list of the tasks, components or steps identified for training from the tasks analysis worksheet. List the components that were identified on the skill-gap analysis worksheet.

ii) Rate each component in terms of trainee's current level of proficiency to perform it on a point continuum.

Frequency with Score

1. Cannot do any part of the task	1
2. Can do less than 50 per cent of the task	2
3. Can complete more than 50 per cent of the task	3
4. Can do the entire task	4
5. Can do the task within the set time	5

iii) Summarize individual responses by adding up the total score for each item and dividing by the number of different forms that were completed, to find out the average proficiency scale.

iv) Identify the components where proficiency is the highest, and where it is the lowest.

v) Review the proficiency ratings and check those tasks that appear to have low proficiency.

vi) Decide whether the gap can be decreased or removed through training.

vii) Discuss the results of the skill-gap analysis with key personnel working in the area and finalize the curriculum.

D. Need Survey

Respondents are asked to provide information about their job title, training frequency, preferred training method, and training material format. Additionally, respondents are asked to rate the effectiveness of the training programs they have participated in and provide suggestions for improvement.

Five Steps: The processes of Training Needs Assessment can be divided into five steps: i) identify problem and needs; ii) determine design of needs assessment; iii) collect data; iv) analyse data; and v) provide feedback.

Methods Used in Need Survey

Advantages and Disadvantages

Advantages	*Disadvantages*
1. Survey a large number of employees at the same time. 2. Do not require a lot of time. 3. Enable honest and open feedback. 4. Gathers quantitative and qualitative data easily.	1. May be difficult to design questionnaires to allow for follow-up or more elaborate responses. 2. Might not identify the specific causes behind employee actions/ behaviours

Personal Interviews

Conducted by a trained "interviewer" who follows an interview outline (or set of questions) to be asked during the interview.

Advantages	*Disadvantages*
1. More flexible in the ability to ask various questions. 2. Able to immediately follow-up on items mentioned in the interview. 3. Is not limited in scope or limited to only a certain set of questions	1. Time consuming. Especially if only one individual is interviewed at a time. 2. Requires the interviewer to document conversations in detail. Any details not documented are lost or need to be gathered through subsequent interviews.

Personal Observations

An observation of the employee at work May be structured (*i.e.*, the employee performs specific tasks) or unstructured (*i.e.*, the observer tries to document the employees work without influencing what the employee does).

Advantages	*Disadvantages*
1. May reduce the amount of interruption of the employee's work. 2. May be more realistic observations are made of the employee actually at work	1. Requires a trained observer. 2. Requires the observer to document the work in detail. Any details not documented are lost or need to be gathered through subsequent observations.

Capacity Development Program at a Glance

Levels of Capacity Development	*Points of Focus*
1. Strategic planning	1. Present or Existing capacity 2. Future Desired Capacity 3. Gap in between existing and desired level 4. Implementation 5. Evaluation
2. Approached of analysis	1. Performance analysis 2. Task analysis 3. Competency analysis 4. Needs survey

Levels of Capacity Development	*Points of Focus*
(i) Performance analysis	1. Definition of Problems 2. Determination of desired output 3. Analysis of performance gap 4. Identification of possible solution 5. Design Learning Solution 6. Evaluation
(ii) Task analysis	1. Identification of task of performance 2. Break up to components 3. Common knowledge analysis 4. On-job training 5. Training topic
(iii) Competency analysis	1. Competency mapping 2. Job description 3. Assessment 4. Critical incident Technique 5. Interviewing 6. Questionnaire 7. Psychometric tests
(iv) Needs Survey	1. Survey questionnaires 2. Personal interview 3. Personal Observation

Chapter 5

Capacity Development Needs Assessment Methods

This Chapter Contains

Sl.No.	Contents
5.1	Data Collection Methods in Identifying Needs
5.2	Rational Methods of Analysis (Observation, Informal Talks, Complaints, Analysis of Report, Opinion Poll, Buzz Session, Analysis of the New Programme
5.3	Information and Skills required in Need Analysis

5.1 Data Collection Methods in Identifying Needs

Data is a collection of facts, figures, objects, symbols, and events from different sources. Organizations collect data using various methods to make better decisions. Without data, it would be difficult for organizations to make appropriate decisions, so data is collected from different audiences at various times.

Data Collection Methods

Data collection methods are techniques and procedures for gathering information for research purposes. They can range from simple self-reported surveys to more complex quantitative or qualitative experiments.

Some common data collection methods include, surveys, interviews, observations, focus groups, experiments, and secondary data analysis. The data collected through these methods can then be analysed to support or refute research hypotheses and draw conclusions about the study's subject matter.

Understanding Data Collection Methods

Data collection methods encompass a variety of techniques and tools for gathering quantitative and qualitative data. These methods are integral to the data collection and ensure accurate and comprehensive data acquisition.

Quantitative data collection methods involve systematic approaches, such as

1. Numerical data,
2. Surveys, polls and
3. Statistical analysis
4. To quantify phenomena and trends.

Importance of Data Collection Methods

Data collection methods play a crucial role in the research process as they determine the quality and accuracy of the data collected

1. **Quality and accuracy:** The choice of data collection technique directly impacts the quality and accuracy of the data obtained. Properly designed methods help ensure that the data collected is error-free and relevant to the research questions.
2. **Relevance, validity and reliability:** Effective data collection methods help ensure that the data collected is relevant to the research objectives, valid (measuring what it intends to measure), and reliable (consistent and reproducible).
3. **Bias reduction and representativeness:** Carefully chosen data collection methods can help minimize biases inherent in the research process, such

as sampling or response bias. They also aid in achieving a representative sample, enhancing the findings' generalizability.

4. **Informed decision making:** Accurate and reliable data collected through appropriate methods provide a solid foundation for making informed decisions based on research findings. This is crucial for both academic research and practical applications in various fields.
5. **Achievement of research objectives:** Data collection methods should align with the research objectives to ensure that the collected data effectively addresses the research questions or hypotheses. Properly collected data facilitates the attainment of these objectives.
6. **Support for validity and reliability:** Validity and reliability are essential to research validity. The choice of data collection methods can either enhance or detract from the validity and reliability of research findings. Therefore, selecting appropriate methods is critical for ensuring the credibility of the research.

The importance of data collection methods cannot be overstated, as they play a key role in the research study's overall success and internal validity.

Types of Data Collection Methods

The choice of data collection method depends on the research question being addressed, the type of data needed, and the resources and time available. Data collection methods can be categorized into primary and secondary methods.

A. Primary Data Collection Methods

(i) Statistical method
(ii) Surveys
(iii) Pools
(iv) Interviews
(v) Delphi technique
(vi) Focus Group
(vii) Questionnaire

1. Primary Data Collection Methods

Primary data is collected from first-hand experience and is not used in the past. The data gathered by primary data collection methods are highly accurate and specific to the research's motive.

Primary data collection methods can be divided into two categories: quantitative and qualitative.

(i) Quantitative Methods

Quantitative techniques for market research and demand forecasting usually use statistical tools. In these techniques, demand is forecasted based on historical data. These methods of primary data collection are generally used to make long-term forecasts. Statistical analysis methods are highly reliable as subjectivity is minimal.

(a) **Time Series Analysis:** A time series refers to a sequential order of values of a variable, known as a trend, at equal time intervals. Using patterns, an organization can predict the demand for its products and services over a projected time period.

(b) **Smoothing Techniques:** Smoothing techniques can be used in cases where the time series lacks significant trends. They eliminate random variation from the historical demand, helping identify patterns and demand levels to estimate future demand.

The most common methods used in smoothing demand forecasting are the simple moving average and weighted moving average methods.

(c) **Barometric Method:** Also known as the leading indicators approach, researchers use this method to speculate future trends based on current developments. When past events are considered to predict future events, they act as leading indicators.

(ii) Qualitative Methods

Qualitative data collection methods are especially useful when historical data is unavailable or when numbers or mathematical calculations are unnecessary.

Qualitative research is closely associated with words, sounds, feelings, emotions, colours, and non-quantifiable elements. These techniques are based on experience, judgment, intuition, conjecture, emotion, *etc.*

Quantitative methods do not provide the motive behind participants' responses, often don't reach underrepresented populations, and require long periods of time to collect the data. Hence, it is best to combine quantitative methods with qualitative methods.

2. Surveys

Surveys collect data from the target audience and gather insights into their preferences, opinions, choices, and feedback related to their products and services. Most survey software offers a wide range of question types.

We can also use a ready-made survey template to save time and effort. Online surveys can be customized to match the business's brand by changing the theme, logo, *etc.* They can be distributed through several channels, such as email, website, offline app, QR code, social media, *etc.*

We can select the channel based on audience's type and source. Once the data is collected, survey software can generate various reports and run analytics algorithms to discover hidden insights.

A survey dashboard can give us statistics related to response rate, completion rate, demographics-based filters, and export and sharing options, *etc.* Integrating survey builders with third-party apps can maximize the effort spent on online real-time data collection.

Practical business intelligence relies on the synergy between analytics and reporting, where analytics uncovers valuable insights, and reporting communicates these findings to stakeholders.

3. Polls

Polls comprise one single or multiple-choice question. They are useful when we need to get a quick pulse of the audience's sentiments. Because they are short, it is easier to get responses from people.

Like surveys, online polls can be embedded into various platforms. Once the respondents answer the question, they can also be shown how their responses compare to others'.

4. Interviews

In face-to-face interviews, the interviewer asks a series of questions to the interviewee in person and notes down responses. If it is not feasible to meet the person, the interviewer can go for a telephone interview.

This form of data collection is suitable for only a few respondents. It is too time-consuming and tedious to repeat the same process if there are many participants.

5. Delphi Technique

In the Delphi method, market experts are provided with the estimates and assumptions of other industry experts' forecasts. Based on this information, experts may reconsider and revise their estimates and assumptions. The consensus of all experts on demand forecasts constitutes the final demand forecast.

6. Focus Groups

Focus groups are one example of qualitative data in education. In a focus group, a small group of people, around 8-10 members, discuss the common areas of the research problem. Each individual provides his or her insights on the issue concerned.

A moderator regulates the discussion among the group members. At the end of the discussion, the group reaches a consensus.

7. Questionnaire

A questionnaire is a printed set of open-ended or closed-ended questions that respondents must answer based on their knowledge and experience with the issue.

The questionnaire is part of the survey, whereas the questionnaire's end goal may or may not be a survey.

B. Secondary Data Collection Methods

Secondary data is data that has been used in the past. The researcher can obtain data from the data sources, both internal and external, to the organizational data.

Internal Sources of Secondary Data	*External Sources of Secondary Data*
Organization's health and safety records	Government reports
Mission and vision statements	Press releases
Financial Statements	Business journals
Magazines	Libraries
Sales Report	Internet
CRM Software	
Executive summaries	

Secondary data collection methods can also involve quantitative and qualitative techniques. Secondary data is easily available, less time-consuming, and inexpensive than primary data. However, the authenticity of the data gathered cannot be verified using these methods.

Steps in the Data Collection Process

The data collection process typically involves several key steps to ensure the accuracy and reliability of the data gathered. These steps provide a structured approach to gathering and analysing data effectively. Here are the key steps in the data collection process:

Sl.No.	*Steps*	*Activities*
1.	Define the Objectives	Clearly outline the goals of the data collection. What questions are to answered?
2.	Identify Data Sources	Determine where the data will come from. This could include surveys, interviews, existing databases, or observational data
3.	Choose Data Collection Methods	Select methods based on objectives and data sources. May be Surveys and questionnaires, Interviews, Focus groups, Observations, document analysis
4.	Develop Data Collection Instruments	Create or adapt tools ensuring valid and reliable

Sl.No.	*Steps*	*Activities*
5.	Select a Sample	Determine how to select sample. Consider sampling methods like random, stratified, or convenience sampling
6.	Collect Data	Execute data collection plan, following ethical guidelines and maintaining data integrity
7.	Store Data	Organize and store collected data ensuring accessibility for analysis
8.	Analyse Data	Now process and analyse it according to objectives, using appropriate statistical or qualitative methods
9.	Interpret Results	Conclude analysis, relating them back to original objectives and research questions
10.	Report Findings	Present findings clearly and organized, using visuals and summaries to communicate insights effectively
11.	Evaluate the Process.	Reflect on the data collection process. Assess what worked well and what could be improved for future studies

5.2 Rational Methods of Analysis (Observation, Informal Talks, Complaints, Analysis of Report, Opinion Poll, Buzz Session, Analysis of the New Programme)

A. Observation

Observation method is described as a method to observe and describe the behaviour of a subject and it involves the basic technique of simply watching the phenomena until some hunch or insight is gained. We are almost constantly engaged in observation.

The 4 main types of observation in sociology are participant observation, non-participant observation, covert observation and overt observation.

The analysis is argumentative and debatable, while the observation gives evidence or reasoning to support the analysis. As we might imagine, the dividing line between what counts as an observation and what is an analysis is not clear-cut.

What is the Observation Method in Research?

The observation method in research involves systematically observing and documenting behaviours, events, or phenomena to gather data. It is a technique used to obtain real-time information about subjects under study. Researchers carefully watch and record what they observe to gain insights and draw conclusions.

The observation method can be applied in various fields, such as social sciences, anthropology, psychology, and more

Steps of Observation Method of Research

The steps of the Observation Method are listed below.

1. **Defining the research objective:** Clearly define the purpose and goals of the observation. Determine what we want to observe and the specific research questions we aim to answer through the observation.
2. **Determining the variables:** Identify the key variables or behaviours we will observe and record. These variables should be relevant to research objective and provide meaningful data for analysis.
3. **Choosing the observation setting:** Select the appropriate setting or environment for the observation. Consider factors such as accessibility, relevance to the research objective, and ethical considerations.
4. **Developing observation protocols:** Establish a structured plan for conducting the observation. Define the procedures, guidelines, and criteria for observing and recording data. This includes specifying the observation techniques, data collection methods, and specific instructions for the observers.
5. **Conducting the observation:** Implement the observation according to the defined protocols. Carefully observe the selected subjects, behaviours, or events while recording the data accurately. Maintain objectivity and avoid any interference or bias during the observation process.
6. **Recording and analysing data:** Record the collected data promptly and systematically. Ensure that the data is organized, labelled, and stored securely. After the observation, analyse the data using appropriate.

B. Informal Talks

In Informal Talks, participants share some of their life experiences and learning experiences when discussing some popular topics, which could help the audience understand their ideas directly.

Talking to people is a constituent element of qualitative research. Researchers are in opinion that conversations are regarded as a key element of social investigation. Researchers maintained that the conversation not only produced valuable data but should be regarded as an important research technique in its own right. Conversations became an integral source of data production in a number of studies by sociologists from the Chicago School.

C. Complaints

By analysing the feedback and complaints from customers, we can identify the root causes of their dissatisfaction, the expectations they have, and the actions we need to take to address them. However, not all complaints are equal, and some may require more attention and resources than others.

Customer complaints analysis is a process used by businesses to evaluate and understand the feedback, especially negative or critical, provided by their customers. This analysis helps companies identify common issues, potential areas for improvement, and opportunities to enhance customer satisfaction

D. Analysis of Report

An analysis report is an essential business report displaying analysis results and feasible suggestions, and providing valuable information for decision-makers so that they can evaluate current operation status and then make well-informed decisions.

What is an Analytical Report?

An analytical report is a written document that businesses use to strategize and make decisions. It examines a specific issue, scenario, or data set, and through analysis, it identifies trends, patterns, and relationships.

In a business context, this report is roadmap for navigating the ever-evolving landscape of market. This type of report can cover anything from financial performance and customer behaviour to market trends and operational efficiency.

Five Steps Writing Analytical Report

1. **Define objective and gather relevant data**
2. **Analyse the data to identify patterns and trends:** To kick-start analysis effectively, consider these tips:
   - (i) Identify key variables
   - (ii) Look for trends
   - (iii) Compare and contrast
   - (iv) Use data visualization
   - (v) Segment data
3. **Structure report for clarity and impact:** Now that we've analysed data and gathered insights, the next step is to structure the report. A well-structured report presents information logically and makes it easy for audience to follow and understand findings.

   Start with an introduction outlining objectives and methodology, then a section presenting data analysis, and conclude with findings and recommendations. When structuring report, consider these tips to enhance clarity and impact:
   - (i) Highlight critical data early
   - (ii) Use clear headings and subheadings
   - (iii) Prioritize data presentation.

(iv) Incorporate visuals
(v) Use bullet points for clarity
(vi) Balance text with white space
(vii) Use a logical flow

The structure of report should guide the reader through thought process, leading them to conclusions coherently and persuasively. Remember, the easier it is to understand report, the more impactful insights will be.

4. **Draft and refine content:** With report's structure in place, the next step is to draft content. This involves transforming data and analysis into a compelling narrative. A well-crafted report communicates the findings and a story that resonates with the audience. Use clear, concise, and jargon-free language to make report accessible to all readers, regardless of their expertise in the subject matter.
5. **Present findings and make recommendations:** The final step is to present findings and recommendations. The presentation should focus on the key insights and recommendations from analysis. Use visual aids like graphs, charts, and info graphics to make complex data more digestible and engaging.

Ending report with solid and actionable recommendations based on analysis is essential. These should be practical and aligned with report's objectives. The goal is to provide value by offering solutions or strategies to help drive growth and efficiency in client's business.

Why is Analytical Reporting so Important?

Analytical reporting is important because it provides the necessary tools to make informed decisions. By highlighting trends and patterns in data, these reports allow businesses to understand their performance and plan strategically.

Analytical reports also make it easier to communicate complex information to stakeholders in a simple, understandable way. This guarantees that everyone, from team members to investors, can make sense of the data and the decisions based on it.

Benefits of Analytical Reporting

Analytical reporting offers several key benefits that can transform how business operates and makes decisions. From better money management to streamlined bookkeeping, these reports make every aspect of running a business more efficient and effective.

1. Improved decision-making
2. A deeper understanding of performance
3. Identification of trends and
4. Increased efficiency.

5. Better communication
6. Strategic planning and Risk management
7. Facilitating growth

Tips to Simplify Analytical Reporting

Analytical reporting can be streamlined with a few innovative strategies. Here are some tips to make the process more efficient and less overwhelming:

1. **Utilize digital tools:** Use tools like expensively to automate data collection and organization, reducing manual effort and potential errors.
2. **Collaborate and get feedback:** Work with team members or stakeholders to review and provide feedback on report. Different perspectives can offer new insights and improve the overall quality.
3. **Focus on storytelling:** Elevate report by weaving data into a compelling narrative. Show how the data tells a story about business, market trends, or customer behaviours.
4. **Conduct situational analysis:** Compare findings with industry benchmarks or competitors through a situational analysis. This context adds depth to analysis and can uncover unique insights.
5. **Emphasize actionable insights:** Ensure each section of report provides actionable insights. Clearly state how these insights can be implemented in business strategies.
6. **Optimize for different mediums:** Consider how report will be consumed (*e.g.*, presentation, PDF, online). Tailor the format and design to suit the medium for maximum impact.
7. **Leverage templates:** Use pre-designed templates, like ours included above, to save time and ensure a consistent format.

E. Opinion Poll

Opinion polls are usually designed to represent the opinions of a population by conducting a series of questions and then extrapolating generalities in ratio or within confidence intervals. A person who conducts polls is referred to as a pollster.

Opinion polls are survey researches to determine voter preference among the candidates running for office and predict the outcome of elections. Exit poll is a post-voting poll.

Expert Opinion Method of Demand Forecasting

The expert opinion method of demand forecasting involves gathering opinions and insights from individuals with expertise in a particular product or market to make predictions about future demand. Conducted just after a voter walks out after casting his or her vote.

Opinion polls or surveys are widely used to study the behaviour of individuals. Survey researchers select and ask questions concerning individuals' attitudes, behaviour, personal characteristics, or other types of information that they can provide as respondents. These responses are counted, analysed statistically, and evaluated or interpreted in different ways, depending on the types of questions asked.

Opinion polls are carried out among a sub-set of a given target audience or population and this sub-set is called a sample.

1. Whilst the number included in a sample may differ, opinion poll samples are typically between c. 1,000 and 2,000 participants.
2. When a sample is selected from a clearly defined target audience or population, the possibility of a sampling error is introduced. This is because the demographic profile of the sub-sample selected may not be identical to the profile of the target audience/population

How are Samples Representative?

1. Opinion polls try, as best they can, to be as representative as possible of a target audience/population.
2. The extent to which samples are truly representative of a given target audience or population is based upon the attention and preparation that has gone into the set up and design of the opinion poll, and the care and attention that has been placed on the sample design and selection.
3. A sample which is not truly representative of a target audience/population is sometimes described as being biased and therefore not representative.
4. There are confidence levels that apply to samples, this is the probability that the sample accurately reflect opinions of that target audience/ population.
5. As only a sample of a target audience/population is typically researched, particular levels of significance apply to the results of opinion polls and these reflect probabilities that opinion poll findings are a true reflection of the target audience/population characteristics being sampled.
6. The confidence intervals are also reflective of the degree of approximation in opinion poll findings. Confidence intervals are an attempt to place mathematical limits to any error which may have occurred in a sample for the opinion poll.
7. There are a number of different ways in which participants can be selected for an opinion poll sample and this depends on the target audience/ population under scrutiny, the ease with which those in the target audience/population can be identified, selected and interviewed as well as the timing and budget available for the opinion poll.

8. Broadly speaking, there are typically two types of sample selection used for opinion polls: Probability and Non-probability samples.
9. Probability samples: Probability samples are also known as "random samples", and this is a method of selecting a sample whereby every member of a target audience/population has an equal chance of being selected to the sample.

Non-probability samples: May be snow ball technique or quota sampling.

How are Opinion Polls Conducted?

1. There are a variety of ways in which research participants can be interviewed either interviewer administered (face-to-face or telephone with interviewers asking questions) or self-completion (an online or postal questionnaire where no interviewers are involved).
2. Research participants are asked (if interviewer administered) or presented (if self-completion) a series of questions contained within a questionnaire. It is important that questions asked as part of opinion polls are credible and fit for the purpose intended any questions asked need to be accurate, balanced, and unambiguous and must not lead research participants to a particular point of view.
3. Once fieldwork is complete and any resultant data analysed results from opinion polls should be reported in an accurate and balanced way, to ensure that any reporting is a true reflection of opinion polling results.

F. Buzz Session

The buzz group method involves breaking participants into small groups of 2-3 people to discuss a specific question or issue for 5-15 minutes. Since the small groups produce "buzzing" sounds while discussing, it is called a buzz group.

Features

1. Small group of 2-3 participants
2. No need to go any other place
3. Less time consuming (5-15min)
4. Discussion only one issue/question
5. Can be used as ice breaker
6. Can be used as supplementing to other methods

Process

1. Decide appropriateness of the method
2. Prepare specific issue/question for discussion
3. Prepare key learning point on the issues

4. Present issue/question
5. Ask individual participants to write down their views
6. Form group of 2-3 participants to write down their common views
7. Allocate 5-15 min for discussion
8. Have them share and discusses issues in the group
9. Supervise each group and encourage individual to participate
10. Stop discussion after expire of time
11. Ask each group to share their points perfectly, one point each group at a time
12. Record them on board
13. Discuss on points and conclude
14. Relate them to key learning point

G. Analysis of the New Programme

Economic changes, new developments in business and industry, and employment opportunities may suggest possibilities for new occupational programs. The stimulus to consider the development of new programs can come from numerous sources including trainees, faculty, advisory committees, the private sector and other community interests.

The elements of a needs analysis include:

1. **Establish a need for the proposed program:** Address the need and rationale for the establishment of the program. The description should include an assessment of workforce demand as well as occupational titles and employment opportunities.
   a. The survey instrument
   b. Target population
   c. Description of sample
   d. Number and name of businesses involved in the sample
   e. Employment opportunities by job titles including:
      1. Number of job openings (due to growth)
      2. Salary for these employees
2. **Projected Student and Graduate Data:** Provide an explanation of potential and identified sources of trainees for the program. Documentation should demonstrate sufficient student interest to support the program and may include surveys of current trainees and related program majors, secondary trainees, as well as employees seeking advanced training. The data should include projected student enrolment and a projection of the number of graduates for the first five years of the program.

3. **Communication with Business and Industry:** Surveys/focus groups should be conducted with local businesses, advisory/planning committees and other workforce training sources such as the Beacon Council or Chambers of Commerce. Community resources should be identified.

5.3 Information and Skills Required in Need Analysis

A needs analysis in this framework is, in essence, a systematic examination of the knowledge, skills, and competencies required by employees and how they align with the organization's goals.

Information needs analysis technique is used to identify all of the information necessary for use in achieving specified goals or objectives, such as performing an activity, satisfying customer needs, or making strategic decisions.

Training needs analysis is assessing which type and level of training are necessary. There are three levels of training needs analysis: organizational analysis, operational analysis, and individual analysis.

A skill gap analysis is a method to identify the gaps in the skills in the organization and take relevant actions to bridge theme technique to improve the company's performance and stay competitive in the long run.

This is an analysis of the job and the requirements for performing the work. Also known as a task analysis or job analysis, this analysis seeks to specify the main duties and skill level required. This helps ensure that the training which is developed will include relevant links to the content of the job.

Chapter 6

Capacity Development Institutions and Management

This Chapter Contains

Sl.No.	*Content*
6.1	Capacity Developer (Trainer) Meaning and Concept
6.2	Trainer: Meaning, Types and Functions
6.3	Good Trainer: Qualities, Skills and Roles
6.4	Training Centres and Training Locations
6.5	Training Model
6.6	Organization's Role in Training

6.1 Capacity Developer (Trainer) Meaning and Concept

Capacity development involves learning and various types of training, along with continuous efforts to develop institutions, political awareness, financial resources, technology systems and the wider social and cultural enabling environment.

An essential feature of training of development professionals is a common understanding of what 'training' and 'training and development' mean.

Deb has summed up the training definition with the following three conclusions.

1. Training is a systematic and planned process to improve the knowledge, skills and attitude of employees.
2. Training provides a learning experience to employees to develop their abilities and improve job performance.
3. Training helps an organization with human resource planning by making available competent and motivated personnel. Therefore, training is a learning process of the acquisition of new skills, attitudes, and knowledge in the context of preparing for entry into a vocation or improving one's productivity in an organization or enterprises.

6.2 Trainer: Meaning, Types and Functions

Trainer plays an important role in the training. Trainers are not only friend, philosopher and guide of trainees, but also are architects of training programmes. According to Paul Chaudhary "a trainer is any person whose resources contribute to the learning and growth of one or more than one persons". It is very important, therefore, to have a professionally qualified trainer, as his qualities and qualifications will highly influence the training and learning process.

Types of Trainers

Types of Trainer	*Features*
1. Regular Staff	These are regular employees of a training institution and carry out training programmes regularly.
2. Ad hoc Staff	Their appointment is ad hoc in nature and they are appointed for the purpose of handling additional workloads
3. Part-time Staff	They are appointed to handle specific assignments. Their appointment is temporary
4. Consultants	Training institutions appoint consultants to help the regular faculty in the preparation and organization of training programmes.
5. Guest Faculty	The guest faculties are invited by the training institutions to take classes on various specialized areas.

Roles of Trainers

The role of a trainer has many facets, and some of the important ones are given below.

i. The Explainer
ii. The trainer is a clarifier
iii. The trainer is a supporter
iv. The trainer is a confronter
v. The trainer is a role model
vi. The trainer is a linker
vii. The trainer is a motivator
viii. The trainer is a translator/interpreter
ix. Trainer as change agent

i) **Explainer**: At the outset, the role of a trainer must be of an explainer. He, or she, must explain the goal, objectives, and practices of the training programme clearly to the trainees. While doing that he, or she, must promote participation and involvement of trainees.

ii) **Clarifier:** The trainer as a clarifier clarifies the doubts of the trainees. Sometimes, the trainees cannot immediately express their expectation from the training. When an individual trainee, or, a group of trainees, is bogged down and uncertain about how to proceed with group work or an individual assignment, then, the trainer could help them and clarify.

iii) **Supporter**: The training group supports the trainees and clear their difficulties. Trainees sometimes feel shy about expressing their views for fear of a "negative halo effect", or they feel that they may be considered incompetent by their fellow trainees. Here, the duty of a trainer is to act as a supporter and a friend to the trainees.

iv) **Confronter:** The trainer must be a confronter. Confrontations are statements or questions that encourage self-examination.

v) **Role model**: The trainer should be a role model for the trainees. Trainer must be friendly with trainees, freely express deficiencies, and get feedback from the trainees and amend himself or herself.

vi) **Linker:** The trainer establishes links with the trainees and the resource persons, both inside and outside the training institution for the effective launching of training programme.

vii) **Motivator:** The trainer as a motivator creates enthusiasm among the trainees to attentively undergo the training programme. Regardless of what the session is about, the trainer must always try to encourage the participants to listen. The trainer can talk to the group before the session starts to identify their needs.

viii) **Translator or Interpreter:** During the process of training, a trainer performs the role of an interpreter or a translator. The trainer translates the views expressed by the trainees in a regional or local language.

ix **Change agent:** The trainer as a change agent views his or her role as that of preparing the trainees for change, so that he or she is able to cope with the internal as well as external change.

6.3 Good Trainer: Qualities, Skills and Roles

Qualities Required for Good Trainers

1. Believe in and have a passion for what they are teaching
2. Know their material well.
3. Use lots of metaphors explaining the materials.
4. Be willing to share their personal experiences with the topic.
5. Have a good sense of humour about the subject.
6. Not get caught up with one person's issues in the class.
7. Keep the class moving at a fast pace without losing anyone.
8. Respect everyone in the class.
9. Involve trainees in group activities.
10. Appreciate the interventions made by the trainees

Skills Required for a Good Trainer

A trainer is incomplete without the possession of the trainer skills. The following intra-personal and inter-personal skills are required for a trainer

a) **Intra-personal skills:** Are required to improve the knowledge base of the trainer as well as being a means for continuous improvement of the self as an effective trainer. This helps the trainer in language skills, different ways of communicating with their trainees, and thoroughly involved in the training programmes. Intrapersonal communication needs intellect as well as physical and emotional involvement. Some of the attributes of intrapersonal skills of a trainer are given below.

1. Thinks
2. Reads
3. Analyses
4. Innovates
5. Nurtures
6. Encourages
7. Reviews

b) Inter-personal skills: Are the skills required for better communication with the participants. This involves active listening, better management of the interruptions, tolerating and encouraging different views. Some of the attributes of interpersonal skill are given below.

1. Talks audibly
2. Relates
3. Accommodates
4. Inspires
5. Notices
6. Entertains
7. Responds

Functions of a Good Trainer

The trainer is the medium to inspire participants to actively participate in the learning process. The trainer introduces a topic to the trainees and communicates the details of subject or skills by selecting methods which directly involve them. A trainer may be an expert in subjects or having good communication skills and is someone who understands group dynamics in training sessions that involve participants with diverse backgrounds. When a trainer with a holistic approach can continue to reach out to participants throughout the training program, then, definitely a training program will be a happy training experience for the trainees; and there will be a sense of accomplishment at the end of the day for the trainer as well as of the trainees. The main functions of a good trainer may be summarized in the manner given below.

Sl.No.	*Functions*
1. Manager of Training Development	Planning, organizing, staffing, controlling training and development projects, and linking training and development with other organizational units.
2. Strategist	Developing long range plans in order to accomplish training and development missions
3. Task Analyst	Identifying activities, tasks, sub-tasks, and human resource and support requirements that may be necessary to attain specific results in a job or organizations
4. Media Specialist	Selecting the right software for programmes, for example, audio visual, computer, and other hardware-based technologies for training.
5. Writer	Developing good training materials
6. Marketer	Selling training and development viewpoints, learning packages, programs, and services to target audiences
7. Facilitator	Facilitating the act of translating learning into practice.

Sl.No.	Functions
8. Instructor	Presenting information and directing structured learning experiences so that individuals learn
9. Counsellor	Helping individuals to enhance their personnel competencies and plan development and career actions
10. Transfer Agent	Helping individuals apply new learning to their tasks
11. Evaluator	Identifying the extent of the impact of a program, service or product and calculate the Return on Investment

6.4 Training Centres and Training Locations

Training institutions should take care of the needs of the trainers as well as of trainees. An atmosphere conducive to learning makes the training process effective. The training organisers of an organisation should ensure that the training is planned in such a way that it achieves its objectives and is also cost-effective.

Training Location

Training programmes may be conducted in-house, within an organisation's premises. Some organisations have fully equipped training centres. Some take their participants to tourist locations, away from the training place. This may increase the interests of the participants, but it involves extra time and money. Some factors which help the trainer and the organisation to decide on the location are given below:

i) Number of participants
ii) Size of training room
iii) Suitable chairs and tables
iv) Lighting
v) Background noise
vi) Access
vii) Facilities
viii) Acoustics
ix) Air conditioning
x) Power outlets
xi) Availability of training aids
xii) Public address system
xiii) Storage area

6.5 Training Model

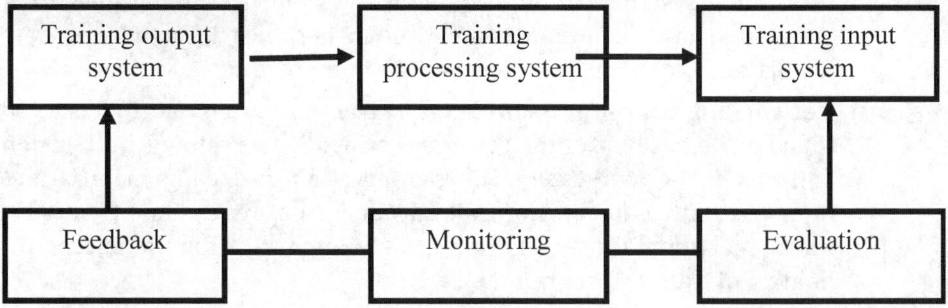

The model starts with training Inputs such as training curriculum, AV aids, resource persons, *etc.* who implement the training or process the training which ultimately result in output such as increase in knowledge, attitude and skills. The feedback is given to training input system, training process is monitored and output is evaluated.

The eight tasks of a Training System, as given by Lynton and Pareek, are discussed below:

i) **Build and maintain support**: The effective implementation of training depends on the development and management of a training system. At the outset, in order to do this it must get support from the top policy and decision makers of the organization.

ii) **Building training competencies and overall training capacity**: The enhancement of competencies of training institutions and trainers is an important goal of training system development. Training institutions must be upgraded with new technologies from time to time, to deal with the changing scenario. The trainers in the training institutions must be oriented from time to time on recent advancements in training methodologies.

iii) **Developing training materials**: Training institutions must be actively involved in the development of training materials. Nowadays, training pedagogy has changed drastically. Training has become more as hands on practice and participatory than one sided lecture discussion. Use of AV aids and ICT in training has completely changed the infrastructure requirements of training institutions.

iv) **Building internal structure and process and developing policy**: Every training institution needs to have a training policy which should clearly delineate its mission, vision, mission, objectives, and activities. Further, the training institutions need to have a work culture and routinely monitor its activities, from time to time.

v) **Strategic planning:** Long term planning should be one of the concerns of the training system. Without systematic and strategic planning, resources are wasted, morale plummets and training becomes the activity of a few idealists, and of many who have failed at other work

vi) **Networking training resources**: The training system motto has to be to make the best use of the resources available for training, both inside and outside the own system. The training system has to find material and human resources needed for training outside its own system. Therefore, intradepartmental coordination is very important for the successful holding of training programme.

vii) **Action spin-offs and boundary management:** According to Lynton and Pareek, training efforts that are well meshed with training needs tend to spark onward actions. These may require further supportive activities on the part of the training system and/or surface related training needs beyond those planned and provided for originally. Such spin-offs are a good indicator of effective training and will bring training more organized support.

viii) **Monitoring and evaluation for appraising and controlling performance**: Monitoring and evaluation is an intrinsic component of a training system. Any task performed by the training system has a purpose. The performance can be judged by monitoring the input and processes and by evaluating the output and outcome attained. Regular monitoring would enable the training system to take action from time to time at various levels.

6.6 Organization's Role in Training

If an organization sponsors a training programme or has requested a customized programme, it has a lot at stake. By sponsoring a training programme, it demonstrates commitment to the training process and participates deeply in the training. It may be keen to get involved in the planning as well as implementation process. Some of the specific tasks in which an organization gets involved are as follow

i) **Identification of training needs:** Regardless of whose initiative has brought the training programme, an organization is expected to assist the training agency in identifying the needs of the participants that it nominates for the programme.

ii) **Formulation of programme contents:** After we have established programme objectives on the basis of the training needs, deducing programme contents from them is relatively easy. Our task is to closely analyse the objectives, interpret them and translate them into training areas and topics.

iii) **Break up programme contents into sub-topics or modules:** This step involves elaborating upon and expressing the contents in distinctly identifiable sub-topics or modules, all set for delivery. It entails breaking up the whole into smaller, recognizable, distinguishable parts. Sometimes, it is difficult to comprehend or understand a training area or topic but when it is broken into smaller constituents, it acquires meaning and relevance.

iv) **Categorise contents into information, knowledge, skills and attitudes:** The next task for us is to categorise the programme contents into components like information, knowledge, competencies, attitudes and values.

v) **Organise and sequence programme contents:** Organising and sequencing the topics and other training activities is a critical element in preparing the programme. These are not put together in an ad hoc or random manner. We should establish a clear linkage, a relationship between various topics and training activities.

vi) **Balance training activities in terms of intensity of participation:** A training activity, sometimes, demands concentrated and intense participation from members of the group. It can extract a lot of their physical as well as mental energy. It is, therefore, recommended that we should sequence the programme in such a way that an intense activity is followed by one that is lighter, and allows the participants to recoup and regain their vigour and enthusiasm.

vii) **Move from simple to complex:** Undeniably, people find it easier to understand and comprehend simple concepts, principles and situations than the more complex and difficult ones. We should, therefore, start the programme with uncomplicated topics and issues that the participants can easily grasp and then, gradually, take them to more complicated ones.

viii) **Move from the known to the unknown:** The process of learning moves faster and the participants learn with minimum effort if we are in a position to consciously relate new learning to what they already know or can do. This movement from the known to the unknown is both natural and logical

ix) **Move from general to specific:** The programme should be so organized that it moves from fundamental or generalized concepts to specifics. Generic consideration of the concepts helps in establishing the basis and framework for discussion of the specifics. This should be a logical course as specifics are derived from generics. This is also a learner-friendly approach.

x) **Maintain the momentum of the programme:** The need to maintain momentum in the programme cannot be overemphasized. The pace should not be allowed to slacken. A lot depends on the manner in which training activities are organized and sequenced during the programme period. We should also give proper attention to a day's schedule.

xi) **Maintain balance between group activities and individual assignments:** Group activity is an important source of learning in any training situation. It provides opportunities to the participants to share their experiences relevant to the topic under consideration of the group.

xii) **Provide opportunities for recap and reflection:** When the participants are engaged in training activities, the entire focus is on the tasks before them. They can be so involved in the activity that they become oblivious to the learning outcomes that the activity aims to realize. If there is a series of such activities, the participants have little time to reflect upon and appraise the benefits that have flowed to them from these activities. They are not able to realize what an activity means as far as their learning and growth is concerned. We should, therefore, prepare the programme schedule in a way that the participants get an opportunity for a recap of the preceding event or activity.

xiii) **Establish common terminology language:** It is important to develop a uniform understanding of concepts, terminology, definitions and nomenclature that are likely to be used more frequently during the programme. This should be done in the initial stage of the programme.

Chapter 7

Capacity Development Project Formulation

This Chapter Contains

Sl.No.	*Content*
7.1	Project Proposal - Concept and Meaning
7.2	Steps in Project Formulation
7.3	Format for Writing Project Proposal
7.4	The Activity Plan

7.1 Project Proposal - Concept and Meaning

A project proposal is a written document outlining everything stakeholders should know about a project, including the timeline, budget, objectives, and goals. The project proposal should summarizer project and sell idea so that stakeholders feel inclined to get involved in the initiative.

Definition of Project Proposal

A project proposal is a document that describes a proposed project and its purpose, outcomes, and the steps that will be taken to complete the project. Not to be confused with a project contract, which is the formal agreement between two parties to complete a project and its purpose is to communicate how a company, team, or individual plan to approach a project.

According to Connor *et al.* (1996), "a proposal is a request for action. Its purpose is to persuade decision makers that a need exists for action and that the action described in the document is the best response to that need". According ITTO, "a project proposal is a detailed description of a series of activities aimed at solving a certain problem. The proposal should contain a detailed explanation of the Project Formulation.

i. Justification of the project
ii. Activities and implementation time line
iii. Methodology, and
iv. Human, material and financial resources required.

A project is a time-bound intervention consisting of a set of planned and interrelated activities executed to bring about a beneficial change. It has a start and a finish, involves a multidisciplinary team collaborating to implement activities within constraints of cost, time and quality and has a scope of work that is unique and subject to uncertainty

Steps in Pre-Project Formulation

The formulation of a good project proposal is not an easy task. It requires a lot of exercise on the part of proposal formulator both before and during the preparation of project proposal. Before writing a project proposal, the project coordination or institution has to take care of following pre-project formulation aspects.

1. **Review of past project proposals:** A group that is involved in the formulation of project proposal needs to review similar types of project proposal formulated by its own institution or other institutions. This will give an idea about the strengths and weaknesses to the project coordinator while thinking about formulating any project.

2. **Consulting experts:** A person or group formulating proposal could consult an expert in the area in which the intended project is going to be formulated and even can appoint a consultant who could be helpful in the preparation of the proposal. It is always better to consult a person who has already completed similar type of projects which are being attempted. The project coordinator formulating project proposal can consult his/her fellow colleagues who have already formulated similar types of project proposals.
3. **Review past project evaluation reports:** Before formulating a proposal, it is advisable to go through the reports prepared by a similar type of research organizations/institutions. The project evaluation report, besides, providing the components of project activities and strategies, will give details about the methodology, evaluation, and impact assessment strategy.
4. **Interact with the prospective beneficiaries:** The project team can also interact with the prospective beneficiaries to be benefited from the project interventions and assess their need. It would be better if the coordinator could also interact with those who have already received benefits from the similar types of project.
5. **Check statistical data/report:** The data regarding a previous similar types of projects from various documents must be collected so that an appropriate project strategy is formulated.
6 **Hold focus group discussion:** It is always better that the person who prepares a proposal undertakes a focus group discussion with the beneficiaries or the prospective clienteles or the stakeholders. If it is a training project for grassroots level representatives *e.g.*, urban local bodies, then the training organizer could conduct a focus group discussion with the elected representatives and functionaries of urban local bodies and assess their needs.

While the actual elements of a project proposal differ depending on the project itself, there are several key elements of a project proposal, which include the followings.

1. Project Background	What problems, challenges, or opportunities exist that create a need for this project?
2. Objectives	What are the intended outcomes of this project?
3. Project Scope	What are the steps or stages of the project? What elements are included in this project? How will the objectives be reached through this project?

Why is a Project Proposal Important?

1. **Improved vendor comparison:** One of the most important purposes of a project proposal is to allow decision-makers to more objectively compare vendors. With a project proposal, decision-makers can compare prices, processes, and projected outcomes. This detailed information helps decision-makers make more informed and objective vendor decision
2. **Project understanding:** A project proposal outlines the purpose and scope of a project. This is helpful before a project takes place since it ensures both parties agree about what the project itself will include. It states to the decision-maker or stakeholder that the service provider understands the scope of the project. This helps establish trust between the parties to facilitate a transactional agreement.
3. **Establish credibility:** A project proposal should be seen as a sales tool that helps establish credibility. It should communicate to the decision-maker or stakeholder that the service provider is capable of completing the project and fulfilling expectations.
4. **Propose an estimated timeline and budget:** A project proposal also proposes a timeline and budget for the project. While this document is not a formal contract, it provides stakeholders enough information about budget, timeline, and outcomes to determine whether or not to move forward with a project and, if so, to make an informed contractor or vendor decision.
5. **Set expectations:** It sets expectations for the client relationship. This includes having a mutual understanding of deliverables, timelines, and budget. While the project proposal is not a contractual agreement, it does set a preliminary agreement that the details in the project proposal will be put into the contract if the project moves forward.
6. **Inform project planning:** Finally, a project proposal also helps to inform project planning. Once the project is approved, this document will be used to finalize timelines and budget for the contract as well as to complete resource scheduling.

7.2 Steps in Project Formulation

Project formulation exercise normally includes the following aspects;

1. Feasibility analysis
2. Techno-economic analysis
3. Project design and Network analysis
4. Input analysis
5. Financial analysis
6. Social cost benefits analysis

7. Pre-investment appraisal
8. Formulation

Stages in Project Formulation

1. Feasibility Analysis

i. First stage in project formulation
ii. Examination to see whether to go in for a detailed investment proposal or not
iii. Screening for internal and external constraints
iv. The project idea seems to be feasible
v. The project idea is not a feasible one
vi. Unable to arrive at a conclusion for want of adequate data

2. Techno-economic Analysis

i. Choice of Optimal Technology, Plan and Design *etc.*
ii. Specifications and standards
iii. Demand for the project output(goods/services)
iv. Overall benefits
v. Provides platform for preparation of detailed project design

3. Project Design and Network Analysis

i. It is the heart of the project
ii. It defines the sequence of events of the project
iii. Time and resources are allocated for each activity
iv. It is presented in the form of a network drawing/bar chart
v. It helps to identify project inputs, finance needed and cost- benefit profile of the project

4. Input Analysis

i. Its assesses the input requirements during the construction and operation of the project
ii. It defines the inputs required for each activity
iii. Inputs include materials, equipment, machines, software, human resources *etc.*

5. Financial Analysis

i. Involves estimation of the project costs, operating cost and fund requirements

ii. It helps in comparing various project proposals on a common scale
iii. Analytical tools used are discounted cash flow, IRR, cost-volume profit relationship and ratio analysis
iv. Investment decisions involve commitment of resources in future
v. It needs caution and foresight in developing financial forecasts.

6. Cost- Benefit Analysis

i. The overall worth of a project
ii. The project design forms the basis of evaluation
iii. Total costs and benefits there to (Benefits > Costs or B/C>1)

7. Pre-investment Analysis

i. The results obtained in previous stages are consolidated to arrive at clear conclusions.
ii. Helps the project-sponsoring body, the project-implementing body and the consulting agencies to accept/reject the proposal.

7.3 Format for Writing Project Proposal

After completing the ground work, the writing of a project proposal will begin. Here, we will mention the important components of a project proposal.

1. Title Page

The project proposal must start with a title page. The title page needs to clearly spell out the title of the project proposal, the name of the sponsoring agency and the institution submitting the project proposal, and the address of the institution from where the proposal originates.

Example:

Title: Training of Urban Health Workers on National Urban Health Mission

Submitted to: Ministry of Health and Family Welfare, Government of Delhi

Submitted by: National Institute of Urban Affairs, New Delhi

2. Abstract

Many project proposal peer group evaluators may not have the sufficient time to read the entire project proposal. Therefore, it is prudent to give an abstract. The abstract needs to include the following aspects.

(i) The problem statement
(ii) The project objectives

(iii) Implementing organisations
(iv) Key project activities
(v) Expected outputs
(vi) Total project budget

3. Introduction

Briefly introduce the topic of the project, and why a project in the areas is deemed necessary. Introduction must be brief and to the point focusing on the topic on which project is being proposed.

4. Problem Statement

It is the statement of situation, objective, problem and solution The project formulator must clearly give some background of the problem which the project is trying to solve in order to Project Formulation.

Problem Analysis

i. Describe the key problem identified and its causes and effects.
ii. Delineate how these problems affect the target group.
iii. Explain how addressing the causes will lead to the eradication of the key problem.
iv. The description should be clear, concise and convincing.

5. Project Goal and Objectives

After delineating the statement of the problem, the second aspect to be covered in the project proposal is to clearly state the goal and objectives. The goal is the general aim of the project, while objectives are the core problems/issues the project is trying to address to achieve the goal. There is one goal and more than one objective. The guidelines for preparing realistic objectives are given below. Guideline for Objectives

Formulation In nutshell objective must be 'SMART'.

S – Simple: It should be simple and covering one point at a time.

M– Measurable: It must be measurable so that a conclusion can be drawn

A – Attainable: It can be handled within the limit imposed by the time

R – Realistic: It must not be unrealistic which cannot be achieved

T – Time-bound: It must be so framed that it is achieved with the set time.

6. Project Output

The project output should describe the services or products which the coordination intend to deliver to the beneficiaries or the survey which the organizers/coordinator intends to administer over the respondents. The results needed to be more detail and can be presented in measurable terms. It should be remembered that the results should address the main causes of the problem that the target group faces.

The following are key points to be taken into consideration while formulating outputs.

(i) It should clearly relate to the objectives and should be stated in such a way that these can be measured in quantity, time, and space.

(ii) It should be achievable within the available resources.

(iii) It should not be confused with activities.

7. Target Group

The project proposal needs to clearly define the target group and show how they will benefit from the project/programme. The project should give clear details of the designation, caste, and age; the gender breaks up of the target group which will be covered in the project. If possible, a list of categories of participants may be given as an annexure in the proposal.

8. Strategies

The project proposal must clearly delineate strategies to be adopted during various steps of project implementation. The formulated strategies must be logically linked with the formulated objectives.

9. Project Implementation Plan

The project implementation plan must describe activities and resources allocation in as much detail as possible. The time and place of project implementation must be clearly mentioned. It should also specify the experts and other institutions who will be involved during the process of conducting the project activities. The implementation plan may be divided into the following two key elements.

7.4 The Activity Plan

The activity plan must clearly include information about the activities to be conducted by the project. The activities must be drawn from the specific objectives and outputs formulated for the project. They should be stated in such a way that they precisely describe the actions or tasks to be undertaken by the project coordinator. It will be better if activities are further broken into tasks. Activities should indicate what will be done to transform inputs into outputs. Some of the guidelines for the formulation of appropriate activities are:

i) It should examine each output listed in the implementation plan.

ii) Formulate the complete set of activities required for achieving the stated objectives.

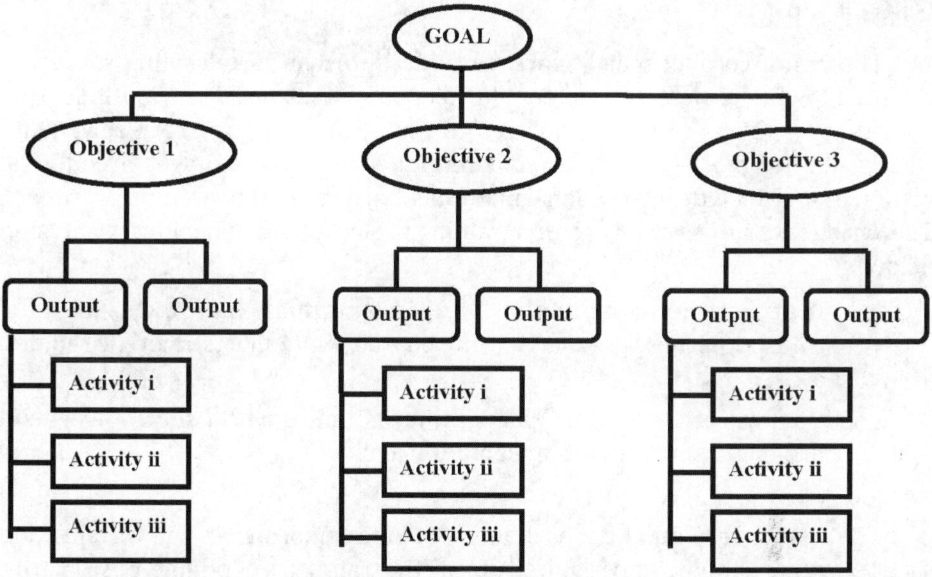

iii) Always verify that the formulated set of activities is sufficient for achieving the stated project output.

iv) Check the available set of inputs necessary for the implementation of formulated activities.

A good tool for establishing activities is the work breakdown structure, which identifies groups of activities related to each project output and presents them in a hierarchical structure as given in flowchart.

The activity map can be presented in a tabular form, popularly known as Gantt chart. An example of training project activities for urban health workers of a training programme for four months is presented in a Gantt chart given below:

Gantt Chart

Sl.No.	*Activities*	*Month-1*	*Month-2*	*Month-3*	*Month-4*
1.	Conducting a need assessment	√			
2.	Preparation of training module	√			
3.	Conducting the training programme	√	√		
4.	Writing Training Report		√	√	√

Resource Plan

The resource plan must clearly provide information regarding resources required for the project. International project donors expect contributions from the project applicant organization. In this case, the training organizer should clearly mention the organization's contribution in the project budget. Broadly, the resources may include instructional materials, hardware, software, and personnel. The various resources required for a training project of urban health are narrated below:

1. Instructional materials: Training manual, handouts, models, transparencies, flow charts, white boards, white sheets, reference guides, *etc.*
2. Hardware: Computer, Video cameras, multimedia setup, overhead projector, power point presentation, *etc.*
3. Software
4. Personnel: Training consultants, resource persons, training assistants, and other support staff required during the training programme. Customarily, the budget for a project proposal for training includes expenditures on following broad headings.
   a) Payment to training coordinators
   b) Payment to training Associate/Assistant
   c) Payment to consultant trainers
   d) Payment to resource persons
   e) Per Diem/TA/DA of the participant
   f) Cost of training materials/modules
   g) Charges of training hall (if available with the organization)
   h) Charges of board/lodging of the participants
   i) Cost of stationary/software/hardware
   j) Miscellaneous charges (postage, photocopying, *etc.*)

Chapter 8

Capacity Development Methods and Tools

This Chapter Contains

8.1 Capacity Development Methods

There are a number of methods to train participants depending on situation, needs of the trainees and objectives of the sponsoring agencies. The following capacity Development Methods are mostly adopted.

1. Lecture
2. Discussion
3. Syndicate
4. Seminars
5. Conference
6. Symposium
7. Role Play
8. Case study
9. Programmed Instruction
10. T – group
11. Laboratory method

1. Lecture

The lecture method, also known as the Trans missive method, is based on vertical learning, whereby the trainer has all to transmit to the trainees. The trainees are considered to have everything to learn.

Concept of Lecture Method

Teaching by lectures is probably one of the oldest methods used by classroom trainers. As a widely practiced method of teaching, a trainer can reach a large number of trainees at the same time; a large amount of materials can be covered in a short period of time. This is a 'trainer-centred' approach involving largely a one-way form of communication from trainer to trainees.

Advantages and Disadvantages

Advantages	*Disadvantages*
The trainer controls the topic, aims, content, organization, sequence, and rate. Emphasis can be placed where the trainer desires	Some of the trainees may already know the content of the lecture while some may not. This may not give the possible effect to the teaching
The lecture can be used to motivate and increase interest, to clarify and explain, to expand and bring in information not available to the trainees,	Lectures are group based. It is possible that the trainer may not be able to pay attention to an individual.

Advantages	*Disadvantages*
Trainees can interrupt for clarification or more detail	The communication is mostly one-way from the trainer to the trainees.
The lecture can be taped, filmed, or printed for future use.	Most of the trainees are not habituated of taking notes.
The lecture can be easily revised and updated.	Lecture information is forgotten quickly, during and after the lecture.
The trainer can serve as a model in showing how to deal with issues and problems.	There is no immediate and direct check of whether learning has taken place.
It is relatively less expensive as no special apparatus is needed.	Lectures are not effective when teaching objectives are not clear
Lecture method channels the thinking of all trainees in a given direction.	Trainees are not very active when only listening.
Large materials can be covered in a short time period.	Few trainers have been taught how to lecture effectively

Guidelines for Effective Use of Lecture Method

1. The trainer should use sufficient teaching aids, good illustration and demonstration while using lecture method to achieve his objectives. The trainees must sit on comfortable chairs/benches, facing away from windows to avoid light rays in their eyes.
2. The trainers should keep distracting noise to a minimum.
3. The room should neither be too cool nor hot. If the trainees are uncomfortable they will be irritated and will not be able to concentrate on what the trainer is saying.
4. The trainer should avoid many movements because it attracts trainees' attention. He should ensure that every trainee sees and hears him from any angle.
5. Trainers should avoid coverage of too many concepts for this may tend to confuse the trainees,
6. Trainers must encourage trainees to ask questions and make comments, as this may reduce boredom.
7. No single teaching method should be used. To maximize learning therefore, a lecture should be followed by discussion, questioning, practice or some other methods.

2. Discussion

The term discussion is derived from the Latin word 'discutere' which means to shake or strike. Discussion means an exchange of ideas between different people or groups of people. In this method of teaching, two-way communication takes place, *i.e.*, both trainer and trainees share their views on a particular concept.

Discussion is important to learning in all disciplines because it helps trainees to process information rather than simply receive it

Steps for Effective Discussion

Steps	*Narration*
1. Preparing for Discussions	To start planning a discussion and decide what want our trainees to get out of the discussion
2. Develop a Clear Goal for the Discussion	We have to decide what learners are to learn. It should be very clear and accordingly the discussion should start.
3. Problematize the Topic	Having a clear goal in mind makes it much easier to plan a discussion. We know what our trainees want to get out of it
4. Select a Discussion Format	Many discussion activities can be used in the classroom. Choose one that will help trainees meet goals for the discussion.
5. Choose a Method to Assign Trainees to Groups	When assigning trainees to groups, consider the following questions. (i) How big should the groups be: Two to six is ideal. (ii) Randomly assigning trainees to groups avoids the problem of friends wanting to get off track (iii) How long should the groups meet
6. Choose a Debriefing Method	Always debrief trainees; it is the most important part of a discussion, the time to summarize and synthesize. Most of learning in discussions happens during debriefing.
7. Problems with Discussion	(i) Getting Started: Trainees are often reluctant to get down to work in a discussion. (ii) Attendance: Trainees may believe that their attendance is not mandatory. (iii) Losing Control: One fear about discussion is the possibility that the discussion will be too enthusiastic or not remain civil. (iv) Discussion Monopolisers: Some participants tend to speak without allowing others

Advantages and Disadvantages:

Advantages	*Disadvantages*
It is a participation oriented method and goes very well with the trainees irrespective of their age.	It is time consuming process.
Synergy takes place and powerful ideas and solutions emerge.	The outcome and decisions of discussions may be controversial.
It is a self-learning methodology, and participants need not depend solely upon the trainers for learning.	Some trainees may dominate the discussion, while others are silent throughout a discussion.
The learning is spontaneous and almost automatic.	
The trainees get on-the-spot feedback from the co-trainees about their opinions and experiences.	

3. Syndicate

Syndicate method is a participative method which basically consists of dividing up the participants into number of groups for the purpose of discussion, in exercise, or work on a project. To undertake a practical exercise or a project.

It is a widely known and popular method applicable to all groups of trainees, especially in giving training to supervisors, managers, and executives. In this method, the trainees are divided into sub groups, called as syndicates and each syndicate has to work on various subjects. The syndicates discuss the issues related to the subject and prepare a report. A syndicate can be formed for any topic. The topic could be one that involves all the departments in an organization, and the members of the syndicate are selected from a cross-section of organizations.

Advantages and Disadvantages

Advantages	*Disadvantages*
All the participants are actively involved in the syndicate sessions. It cultivates leadership qualities since the leadership roles president, secretary, *etc.* - are filled by rotation.	If the syndicate consists of members with wider interests and views, there may be disagreements on the issue till the end.
It is a self-learning and development process because each of the syndicate members comes out with his knowledge, experiences, and skills	Hurriedly structured syndicates do not contribute anything.

Advantages	*Disadvantages*
Generally, the issue is discussed in detail and the trainees look at the problem from many	
The syndicates are formed for certain definite purpose.	
The existence of the syndicate is for some fixed duration and there is consistent interaction and understanding between trainees.	

4. Seminars

Seminars involve a group coming together to discuss and learn about specific topics. It is a teaching technique where guided interaction among the group helps deeper understanding of the subject matter.

Seminars can be classified based on size into mini, major, national and international. Seminars are short term training programmes arranged for a few days may be a day or two or maximum three days

Advantages and Disadvantages

Advantages	*Disadvantages*
This method helps participants to obtain the views, opinions, research finding, *etc.* of the experts through the papers presented by them	If the audience for the seminar is too big, individual attention, is not possible. seminars
The method is faster than gathering knowledge by reading books and holding discussions	Many seminars have just become rituals.
Some seminars are forceful and thought provoking. They enable the participants to see different points of view, attitudes, and opinions	Interactions between participants are very limited.
In seminars, the trainees are given the opportunity to listen to experts from different fields who share their views on specific issues	

5. Conference

For conveying this type of training, a conference also known as a meeting is carried out where a certain plan is spoken about to the people and they get their training and knowledge through it. It explains all the necessary facts as well as important principles and concepts needed to be known.

Conferences are suited to executives and senior officers who are dealing with a complex problem which requires contributions from senior officers for a resolution. The issues and the subject matters of the conference are intimated to the delegates and participants well in advance, and they prepare the papers and related materials for the conference. The delegates present the papers and explain the various aspects of the problem. At the end of the conference, suggestions and recommendations are made, based on the deliberations and discussions. All the members in a conference are expected to have more or less equal knowledge

Advantages and Disadvantages

Advantages	*Disadvantages*
This method suits bigger groups and participants benefit from listening to senior speakers. They are induced to think over the problems, issues and solutions	It is less participatory and the trainees are inactive during the programmes. It is difficult to determine whether the trainees learnt anything or not

6. Symposium

Symposium method consists of having a chairman or symposium leader, plus a group of, usually, four, five, or six persons, each of whom presents, in a concise, organized, more or less formal way, one phase of the main topic selected for discussion.

In a symposium, a particular subject is discussed widely and deeply from the viewpoint of several well-versed authors. The speakers are of approximately equal ability, to avoid one speaker dominating the meeting. The participation of the trainees is more rigorous and learning more thorough

Advantages and Disadvantages

Advantages	*Disadvantages*
The main advantage of this method is the active thinking and participation by participants.	A lot of controversy and confusion in thinking may result in futile endeavour.

Differences between Symposiums and Seminars

The main difference between a symposium and a seminar lies in their overall purpose, format, and scope. Here are the key distinctions:

(i) **Purpose:** A symposium brings together experts to present and discuss a specific topic or theme. It emphasizes the exchange of knowledge, research findings, and insights among participants.

In contrast, a seminar is more focused and interactive, typically used for training or professional development. They encourage skill-building, knowledge transfer, and active engagement among a smaller group of participants.

(ii) **Format:** Symposia often feature a series of presentations, panel discussions, and breakaway meetings. They may span multiple sessions or days, allowing participants to attend the sessions that align with their interests.

Seminars typically involve a single session or a series of closely related meetings, often in smaller groups. They prioritise interaction, active participation, and engagement between the attendees. Seminars often include discussions, question-and-answer sessions, and practical exercises.

(iii) **Expertise:** Symposiums are often championed for their inclusive nature. Companies tend to invite all employees to as they foster interdisciplinary dialogue and collaboration among experts and practitioners.

On the other hand, seminars are structured to suit a specific expertise level. This can be for beginners, intermediate learners, or advanced professionals, providing education and training that aligns with their specific needs.

(iv) **Scope:** Symposiums, whilst undoubtedly having a core topic, attract a more diverse audience. Typically, a wide range of topics are covered that relate to the overarching theme. As seminars target a specific level of expertise the range of topics is often much narrower.

Overall, symposia focus on knowledge exchange and comprehensive discussion on a specific topic, whereas seminars are more intimate meetings designed for skill-building and targeted personal development. Symposia are a solution for broader perspectives whereas seminars are more suited to learning environments where interaction between participants is required.

7. Role Play

Role-play training is a learning technique in which participants act out scenarios under the guidance of a trainer. This type of training can create a safe environment for employees to build confidence in their abilities to succeed in their jobs.

Role play is a form of experiential learning (Russell and Shepherd, 2010). Trainees take on assigned roles and act out those roles through a scripted play.

The role play can be carried out one-to-one (individual role play) or as a group role play with each member in the group taking on a role/character.

Why is Role-Play Training Beneficial?

(i) To promotes understanding about the situation

(ii) To develop empathy; and

(iii) Self-perception is useful in helping individuals develop the art of interpersonal relationship.

Steps to Conduct Role-Play Training

1. **Identify an issue:** Before the role-play session, gather participants and introduce a hypothetical issue they may encounter during their jobs. Make sure participants clearly understand the problem we are trying to address.
2. **Describe a specific scenario:** Choose a scenario for the participants to act out. Describe it thoroughly, providing specific details to make the role-play session as realistic as possible. Consider using scenarios that participants might experience in their work, such as collaborating on team with other employees or marketing a new product
3. **Assign roles:** To the participants roles are to be assigned keeping suitability in view.
4. **Have participants act out the scenario:** Ask some participants to act out the scenario and create different strategies for resolving the situation. For example, if the goal of role-play activity is to have employees practice giving a sales pitch, participants might develop different ways to present a product to prospective customers.

Types of Role Play

Play Role play can be broadly classified into three categories.

(i) Structured/Rehearsed Role Play/Schemed Role Play.

(ii) Unstructured/Spontaneous Role Play.

(iii) Controlled and directed Role Play.

(i) Structured/Rehearsed/Schemed Role Play: Without any scripts, the participants are told in advance about what action, attitude and behaviour each character should act out. It is a lively visual aid used to portray varying situations in the work field

(ii) Unstructured/Spontaneous Role Play: Participants are involved in the role play without any brief or preparation. They act out according to their mind, attitudes, and styles.

(iii) Controlled and Directed Role Play: The trainer gives directions to the participants of role play along with the script. The characters have no choice or freedom except to act according to the wishes of the trainers. They have to memorize the script and present it before the audience. It is an effective tool for the trainer to present to the audience what he wants to say and the topics which he wants to cover.

Advantages and Disadvantages

Advantages

1. The trainees have a feel of a real work life situation. The role play bridges the gap between theory and practice.
2. Trainees have opportunities to develop skills in communication, interpersonal relationships, decision making, observation, *etc.*, through this methodology.
3. It is an active methodology, and can be utilized by trainers during a post lunch session when the trainees feel drowsy and dull.
4. Spontaneous role play helps the trainees to think and analyse the situation independently and logically.
5. There are many constructive criticisms and suggestion which serve as an eye opener for other participants.

Disadvantages

1. The main disadvantage of role play is that it is time consuming and well experienced trainers are needed to conduct role play sessions.

8. Case Study

The case study is a method which provides descriptive situations which stimulate trainees to make decisions. The purpose of the case method is to make trainees apply what they know, develop new ideas to manage a situation or solve a problem.

This method has been introduced by the Harvard Business School and widely used by many of the management institutes in India and abroad

Types of Case Studies

Picking the right type of case study for training programme is important.

Types of case studies are:

1. Descriptive case studies: Focus on explaining a particular situation or action.
2. Exploratory case studies: Explore the potential benefits and limitations of existing strategies or examine emerging trends or new phenomena.
3. Instrumental case studies: Focus on understanding a particular problem and then provide insights into a broader issue or problem.

4. Intrinsic case studies: Focus on a particular case and then generalise the findings to other scenarios.
5. Collective case studies: Examine a group of related cases to gain insights into broader phenomena.

Advantages and Disadvantages

Advantages	*Disadvantages*
It is a lively and interesting method. It keeps the participant vibrant and encourages participation in the learning process.	It is a time consuming and may be cumbersome at times.
It develops the thinking, analysing, and problem solving skills of the participants.	Lack of provision of adequate data and incidents may dilute the case study method.
It relieves monotony of listening to continuous faculty lectures.	There is a shortage of appropriate cases.
Since cases are made out of real life situations, it helps the participants to solve the problems.	If the case studies are lengthy, the participants may lose their concentration.
Relationship building skills are developed.	It requires skill on the part of trainer.

9. Programmed Instruction

Programmed instruction is based on theories of psychology, more specifically operant conditioning. It is a new technique of teaching learning. Programmed instruction is the presentation of material in a step by step procedure with the knowledge of results, and the possibility of different courses through instruction.

Programmed instruction is a method of presenting new subject matter to trainees through a graded sequence of controlled steps with corresponding activities

Assumptions of Programmed Instruction

Programmed instruction is based on the following assumptions:

1. A trainee learns better when he himself is active.
2. A trainee learns better if he is motivated to learn by confirming his responses.
3. A trainee learns better if subject-matter is presented to him in small steps (units).

4. A trainee learns better if sequence of the content is psychologically workable.
5. Learning may be effective if the pre-requisite on the part of the learner are specified.

Principles of Programme Instruction

The basic principles of programmed instruction are as follows:

1. Principle of small steps
2. Principle of active responding
3. Principle of feedback
4. Principle of self-pacing
5. Principle of error control

1. **Principle of small steps:** The subject or skill to be acquired is broken into small steps. Learning takes place most rapidly if the subject matter is presented in small steps gives a feeling of satisfaction which in turn reinforces learning.
2. **Principle of active responding:** Another principle of programmed instruction is based on active responding by the trainees. Learners are made to interact with every bit of information and make a response. Responses may be overt (when learners write the answers) or covert (when learners think out the answers).
3. **Principle of feedback:** The term feedback means 'knowledge of results'. Experimental evidence supports this principle that the more immediate, the reward better is the learning.
4. **Principle of self-pacing:** Programmed instruction is based on the assumption that the individual learner learns according to his own pace, needs and capabilities. It is trainee-centred and encourages each to work at his own speed.
5. **Principle of error control:** Programmed instruction is closely linked to the issue of error control. The learning sequence is broken down into a large number of small steps, so that the rate of error is kept minimum and allows reinforcement to be frequent and immediate.

Characteristics of Programmed Instruction

1. It is a part of educational technology in the sense that programmed material can be presented with the help of machines or computers.
2. It is a new strategy of teaching and learning. Here, learner learns himself without the help of the trainer.
3. It is a technique for the solution of educational as well as teaching problems.

4. It is a technique for the modification of learner's behaviour. Thus, it is opposite of micro teaching where trainer's behaviour is modified
5. It cannot replace the trainer from the field of teaching altogether. It is because only an effective trainer can prepare a good programmed material.
6. It requires more creativity and imaginative efforts to develop such individualized instructional material. Every individual learner learns at his own speed here.

Advantages and Disadvantages

Advantages	*Disadvantages*
This method enables the trainees to learn at their own speed and convenience.	Writing the programme is tedious, time consuming
A large number of people can learn through the programmed learning method without any geographical or distance barriers.	Learning through this method requires strong motivation on the part of the trainees to learn
The infrastructural facilities such as a well-developed faculty, class room, and audio visual aids are not needed.	It is a highly structured routine methodology. Some people may resent directed learning.
This method is highly elastic and flexible. The same programme can be used for many batches of trainees, year after year, and it is easier to introduce modifications in the programmes	
The trainees get the psychological satisfaction due to immediate feedback, self-pacing, and individualized learning.	

10. T-Group

The T(training)-group is a training device. In this group 10-20 people meet with no agenda to explore what happens naturally in a group and how the group tries to solve the problems that evolve. The group usually meets for two-three weeks for several times a day.

The T-group provides participants with an opportunity to learn about themselves, their impact on others and how to function more effectively in group and interpersonal situations.

A T-group or training group (sometimes also referred to as sensitivity-training group, human relations training group or encounter group) is a form of group training where participants (typically between eight and fifteen people) learn about themselves (and about small group processes in general) through their interaction with each other. They use feedback, problem solving, and role play to gain insights into themselves, others, and groups.

11. Laboratory Method

The laboratory method involves hands-on learning through original data and materials. It aims to provide an active learning experience for trainees. The purposes of the laboratory include teaching psychomotor skills, orienting trainees to new equipment and procedures, and supplementing classroom work.

Sensitivity training is called as laboratory training as it is conducted by creating an experimental laboratory situation in which employees are brought together in groups, to interact in an unstructured environment. In sensitivity training social sensitivity and behaviour flexibility is developed within the employee.

Principles of Laboratory Method

1. It follows the principle of learning by doing.
2. It follows psychological principle, where trainee's age and interest is taken into consideration.
3. The work should be Pre-organized and Pre-selected.
4. Trainer must see that, trainees are allowed to work independently without much interference.
5. The trainer must ensure that apparatus and equipment should be checked pair hand.
6. Trainer must see that trainees are able to follow instruction and record their observation properly

Role of Trainer

1. Trainer must be a facilitator of the process of doing experiments by trainees.
2. Trainer must check the apparatus previously, so that it goes on smoothly.
3. The practical work must be Pre-organized and Pre-selected.
4. The skills of handling apparatus, drawing, diagrams, careful observations taking necessary precautions, must be developed among trainees.
5. The trainer must be that, the student is doing experiment properly by following proper procedure.

Advantages and Disadvantages

Advantages	*Disadvantages*
This method follows child-cantered approach.	It is expensive and uneconomical
It makes trainees active and alert.	It is time consuming.
It gives scope for learning by doing and trainees do a lot of thinking themselves.	It expects a lot from trainees and trainer.
Different skills are developed.	It does not guarantee that, trainees would be equally efficient in solving problems outside laboratory.
It paves way for exploration experimentation and verification of scientific facts and principles.	All trainees cannot be expected to be skilled workers.
It inculcates good virtues like, honesty, truthfulness, dignity of labour *etc.*	Most of the trainees are either not ready or lack to ability to undertake original work.
It helps in developing sprit of enquiring.	
It helps in developing higher order like reasoning, analysing, synthesizing *etc.*	

8.2 Capacity Development Objectives and Appropriate Training Methods

1. Training Objectives

Training objectives specify the type of learning outcomes intended from the training programme. If the purpose of training is to increase knowledge, the training methods that allow presenting the concepts with examples in a logical sequence and reviewing key points will be effective. The suitable methods for this purpose are listed below.

To Increase Knowledge	*To Acquire New Skills*	*To Change Attitudes*
1. Reading assignment	1. Demonstration	1. Role play
2. Lecture	2. Role play	2. Case study
3. Audio and video tapes	3. Video tapes	3. Demonstration
4. Case study	4. Structured exercise	4. Field visit
5. Demonstration	5. Workshop	5. Subject Matter
6. Group discussion		
7. Buzz session		
8. Brain storming		

2. Subject Matter

Training methods vary according to subject matter.

Types of Subject Matter	*Methods to be Used*
1. If training is on theoretical or conceptual clarity,	1. Lecture 2. Audio 3. Video tapes and 4. Reading assignments
2. If one right answer which needs minimum debate	1. Lecture 2. Audio or 3. Videotape
3. If debatable outcome is required	1. Group discussion 2. Workshops and 3. Case study

3. Participants

Training methods vary as per participants.

Learning Styles	*Learners' Experience*	*Size of Group*
1. Self- discovery methods	1. Level of qualification	1. Simulation method
2. Independent study	2. Graduate	2. 8-10 participants
3. Clinical experience	3. Under graduate	
4. Practical exercises and	4. Matriculation	
5. Interviewing	5. Little of no qualification	

4. Training Methods and Size of Group

Sl.No.	*Teaching Methods*	*Number of Participants*
1.	Lecture	10-40
2.	Group discussion	3-10
3.	Case study	3-20
4.	Role play	5-15
5.	Field Visit	5-25
6.	Seminar, Syndicate	15-40
7.	Demonstration	5-15

Sl.No.	Teaching Methods	Number of Participants
8.	Simulation method	8-20
9.	Structured experience	8-20
10.	Instrumental experience	8-15

5. Resource Availability

Availability of adequate resources to conduct a training programme must be considered while choosing training methods. The following points should be answered before the selection of method.

(i) Competency of Trainer

(ii) Financial provision

(iii) Availability of facilities.

(iv) Location of training

6. Time

In many cases, the participants have a limited number of days to be spent for the purpose of training.

7. Criteria of Selection of Teaching Method

1. It should increase the interest of the learners.
2. It should encourage the trainees to participate.
3. It should motivate the trainees to study and explore.
4. It should develop the creativity of the trainees.
5. The method should be according to the intelligence level of the trainees.
6. The method should link bookish knowledge with real-life situations.

8. Factors that Determine the Selection of Teaching Methods

1. The ability of the trainer
2. The syllabus and curriculum
3. The size of the class
4. The intelligence level of trainees
5. Availability of teaching aids
6. Age of the learners

Chapter 9

Capacity Development Programme Evaluation

This Chapter Contains

Sl.No.	*Contents*
9.1	Meaning, Importance and Purpose of Evaluation
9.2	Principles of Evaluation
9.3	Types of Evaluation - Formative, Summative
9.4	Types of Assessment
9.5	Models of Evaluation
9.6	Process of Evaluation - Evaluation at the Beginning, During and End of the Program
9.7	Statistical Tools for Evaluation of Training
9.8	Use of Evaluation Findings
9.9	Indicators of Training Effectiveness

9.1 Meaning, Importance and Purpose of Evaluation

A. Meaning of Evaluation

The United Nations Development Program (UNDP), defines CD as "the process through which individuals, organisations, and societies obtain, strengthen, and maintain the capabilities to set and achieve their own development objectives over time.

The Organisation for Economic Coordination and Development (OECD) defines, CD as "the process whereby people, organisations and society as a whole unleash, strengthen, create, adapt and maintain capacity over time."

Global Affairs Canada (formerly Canadian International Development Agency, CIDA) defines CD as "the activities, approaches, strategies, and methodologies which help organizations, groups and individuals to improve their performance, generate development benefits and achieve their objectives."

PACT defines CD as "a continuous process that fosters the abilities and agency of individuals, organizations, and communities to overcome challenges and contribute towards positive social change. Though often developed in response to an immediate and specific issue, capacities are adaptable to future opportunities and challenges."

Training evaluation is the systematic process of analyzing training programs to ensure that it is delivered effectively and efficiently. Training evaluation identifies training gaps and even discovers opportunities for improving training programs.

By collecting feedback, trainers and human resource professionals are able to assess whether training programs are able to achieve its intended outcome, and if the training materials and resources used are aligned with or meet company and industry standards

Benefits

Evaluation of training gives comprehensive feedback on the value of the training programs and their effectiveness in achieving business goals. It helps the management to better understand and identify skill gaps to analyze the desired outcomes of training programs. It also helps the organization to:

i. Identify issues and improve the overall processes of training programs
ii. Analyse the effectiveness of training materials and other tools
iii. Determine the needed leadership competencies to solve critical problems
iv. Support continuous change in career development and
v. Assess the overall training experience of the participants

B. Importance

Evaluation Capacity Development (ECD) efforts strengthen and sustain both individuals and organizations to access, build, and implement evaluative knowledge and skills; Design, implement, and manage effective evaluation processes and findings; and. Use evaluations as a performance improvement tool.

Capacity building encourages individuals to think creatively and develop new ideas and solutions to different challenges. It helps them broaden the scope of their work to serve more people or to serve the people they are already working with efficiently.

C. Purpose of Evaluation

1. Involves continuous or ongoing collection and analysis of information about implementation to review progress;
2. Compares actual progress with what was planned so that adjustments can be made in implementation; and
3. An internal activity that is the responsibility of those who manage implementation procedures, thus representing a good management practice?
4. It is a periodic assessment of the relevance, efficiency, effectiveness, impact, and/or sustainability of an activity or intervention;
5. It is a systematic search for answers about an intervention (project, program, or policy);
6. Measures the effects of an intervention and compares them with the goals and objectives of the intervention;
7. Access, build, and implement evaluative knowledge and skills;
8. Design, implement, and manage effective evaluation processes and findings.

9.2 Principles of Evaluation

Capacity development builds upon respect and self-esteem. There are no blueprints. Capacity development draws upon voluntary learning, with genuine commitment and interest. Knowledge cannot be transferred, it needs to be acquired.

The evaluation process involves planning, method selection, data collection and analysis, reporting, and dissemination. Both qualitative and quantitative methods are used. The goal is to improve programs and determine their effectiveness, efficiency, and relevance in improving health

Ten Principles for Capacity Development

Principle	*Description*
1. Don't rush	Capacity development is a long-term process. It eludes delivery pressures, quick fixes and the search for short-term results.
2. Respect the value system and foster self-esteem	The imposition of alien values can undermine confidence. Capacity development builds upon respect and self-esteem.
3. Scan locally and globally; reinvent locally	There are no blueprints. Capacity development draws upon voluntary learning, with genuine commitment and interest.
4. Challenge mind-sets and power differentials	Capacity development is not power neutral, and challenging mind-sets and vested interests is difficult. Frank dialogue and a collective culture of transparency are essential steps.
5. Think and act in terms of sustainable capacity outcome	Capacity is at the core of development; any course of action needs to promote this end.
6. Establish positive incentives	Motives and incentives need to be aligned with the objective of capacity development, including through governance systems.
7. Integrate external inputs into national priorities, processes and systems	External inputs need to correspond to real demand and be flexible enough to respond to national needs and agendas.
8. Build on existing capacities rather than creating new ones	This implies the primary use of national expertise, res and strengthening of national institutions.
9. Stay engaged under difficult circumstances	The weaker the capacity, the greater the need. Low capacities are not an argument for withdrawal or for driving external agendas.
10. Remain accountable to ultimate beneficiaries	Any government is answerable for its people and should foster transparency as the foremost instrument of public accountability.

Lopes and Theison 2003.

9.3 Types of Evaluation - Formative, Summative

A. Formative Evaluation

Formative evaluation provides ongoing feedback and data to guide the development process and ensure that the final product or service meets the intended goals and objectives. Formative evaluation provides ongoing feedback and data to guide the development process and ensure that the final product or service meets the intended goals and objectives.

B. Summative Evaluation

Summative evaluation provides feedback on the overall effectiveness of the program in achieving its intended goals and objectives. Summative evaluation, can be referred to as ex-post evaluation (meaning after the event), typically occurs after implementation, and focuses on the impact of a given technology. Most often this form of evaluation is used to measure whether or not a technology achieved its desired aims or goals.

To leverage formative and summative evaluation for program success, it is essential to establish clear goals and objectives, use a variety of data sources and research methods, involve stakeholders, ensure data quality, use data to inform decision-making, communicate findings effectively, ensure ethical considerations, and continuously assess and adjust the evaluation plan.

By leveraging the insights gained from formative and summative evaluation, program developers can make data-driven decisions that lead to the success of their programs, products, or services. Therefore, it is essential to use appropriate research methods for each type of evaluation to ensure that the data collected is reliable and valid and can be used to make informed decisions.

9.4 Types of Assessment

There are three types of assessments: diagnostic, formative and summative. Although are three are generally referred to simply as assessment, there are distinct differences between the three.

A. Diagnostic Assessment

Diagnostic assessment can help to identify trainee's current knowledge of a subject, their skill sets and capabilities, and to clarify misconceptions before teaching takes place.

Types of Diagnostic Assessments

i. Pre-tests (on content and abilities)
ii. Self-assessments (identifying skills and competencies)
iii. Discussion board responses (on content-specific prompts)
iv. Interviews (brief, private, 10-minute interview of each student)

B. Formative Assessment

Formative assessment provides feedback and information during the instructional process, while learning is taking place, and while learning is occurring. Formative assessment measures training progress but it can also access own progress of an instructor.

A primary focus of formative assessment is to identify areas that may need improvement.

Types of Formative Assessment

(i) Observations during in-class activities of trainees and non-verbal feedback during lecture

(ii) Homework exercises as review for exams and class discussions)

(iii) Reflections journals that are reviewed periodically during the semester

(iv) Question and answer sessions,

(v) Conferences between the instructor and trainees at various points in the training

(vi) In-class activities where trainees informally present their results

(viii) Trainees feedback collected by periodically answering specific question about the instruction and their self-evaluation of performance and progress

C. Summative Assessment

Summative assessment takes place after the learning has been completed and provides information and feedback that sums up the teaching and learning process. Typically, no more formal learning is taking place at this stage, other than incidental learning which might take place through the completion of projects and assignments.

High-stakes summative assessments typically are given to trainees at the end of a set point during or at the end of the session to assess what has been learned and how well it was learned. Grades are usually an outcome of summative assessment: they indicate whether the one has an acceptable level of knowledge-gain.

Summative assessment is more product-oriented and assesses the final product, whereas formative assessment focuses on the process toward completing the product. Once the project is completed, no further revisions can be made. If, however, trainees are allowed to make revisions, the assessment becomes formative, where trainees can take advantage of the opportunity to improve.

Types of Summative Assessment

i. Examinations (major, high-stakes exams)

ii. Final examination (a truly summative assessment)

iii. Term papers (drafts submitted throughout the semester would be a formative assessment)

iv. Projects (project phases submitted at various completion points could be formatively assessed)

v. Portfolios (could also be assessed during it's development as a formative assessment)

vi. Performances
vii. Student evaluation of the course (teaching effectiveness)
viii. Instructor self-evaluation

Assessment measures if and how students are learning and if the teaching methods are effectively relaying the intended messages. Hanna and Dettmer (2004) suggest that we should strive to develop a range of assessments strategies that match all aspects of their instructional plans. Instead of trying to differentiate between formative and summative assessments it may be more beneficial to begin planning assessment strategies to match instructional goals and objectives at the beginning

9.5 Models of Evaluation

A. Will Thalheimer's Training Evaluation Model

Will Thalheimer's Training Evaluation Model is a renowned model used to evaluate the effectiveness of training programs widely known as Learning-Transfer Evaluation Model (LTEM). Thalheimer's updated model consists of 7 levels of evaluation:

1. Level 0: Did not participate - This level represents those who did not participate in the learning program and serves as a baseline for comparison.
2. Level 1: Attendance - This level measures the number or proportion of learners who attended the learning program.
3. Level 2: Completion - This level measures the number or proportion of learners who completed the learning program.
4. Level 3: Competence - This level measures the degree to which learners have acquired the knowledge, skills, and abilities taught in the learning program, often assessed through tests or other evaluations.
5. Level 4: Performance - This level measures the degree to which learners are able to apply what they learned to their job or other relevant contexts, often assessed through observation or other performance evaluations.
6. Level 5: Transfer - This level measures the degree to which learners are able to apply what they learned in the learning program to their job, and the extent to which the organization supports and reinforces learning transfer.
7. Level 6: Business results - This level measures the impact of the learning program on business outcomes, such as increased productivity or revenue, reduced errors or accidents, or other relevant metrics.

B. The Hamblin-Kirkpatrick Model

The Hamblin-Kirkpatrick model, also known as the Kirkpatrick model, is a widely used framework for evaluating the effectiveness of training programs. The model consists of four levels of evaluation, each building on the previous one, as follows:

1. Level 1: Reaction - This level measures how participants respond to the training. It includes collecting feedback from participants through surveys or interviews to gauge their satisfaction, engagement, and relevance of the training.
2. Level 2: Learning - This level measures the extent to which participants have gained new knowledge, skills, and attitudes as a result of the training. This could include assessing performance through tests, assignments, or other measures.
3. Level 3: Behaviour - This level measures whether participants have applied what they learned in the training to their work. It includes evaluating changes in job performance or behaviour, such as increased productivity, quality of work, or changes in behaviour.
4. Level 4: Results - This level measures the impact of the training on the MDO as a whole. It includes evaluating the business outcomes or benefits of the training, such as increased revenue, reduced costs, or improved customer satisfaction.

The model was developed by Donald Kirkpatrick and later revised by Roger Hamblin. By evaluating training programs at all four levels, institute can gain a comprehensive understanding of the effectiveness of their training initiatives and make informed decisions about future investments in training and development.

C. The CIRO Model

The Content Input Reaction and Output (CIRO) model is a popular approach used for evaluating short term training programs. In the context of civil service officials in India, the CIRO model can be applied as follows:

1. Content: The first step in evaluating the effectiveness of a training program is to assess its content. This involves reviewing the program's objectives, curriculum, and teaching methods to ensure they are aligned with the needs of trainees. It is essential to ensure that the training covers all the critical topics and skills required to perform the duties.
2. Input: The second step is to evaluate the input, which refers to the resources and effort invested in the training program. This includes assessing the qualifications and expertise of the trainers, the quality of training materials, and the facilities and equipment provided for the training. It is also important to ensure that the training is delivered in a conducive and supportive learning environment.

3. Reaction: The third step in the CIRO model is to evaluate the reaction of the trainees who participate in the training. This involves gathering feedback from the trainees to determine their satisfaction with the training program, the relevance of the content, and the effectiveness of the teaching methods. Feedback can be collected through surveys, interviews, or focus group discussions.
4. Output: The final step is to evaluate the output, which refers to the impact of the training program on the performance of trainees. This involves assessing the extent to which the training has improved the skills and knowledge of the officials, and whether it has resulted in improved job performance. This can be measured through performance metrics, such as productivity, efficiency, and effectiveness.

Overall, the CIRO model provides a comprehensive framework for evaluating the effectiveness of short-term training programs. By assessing the content, input, reaction, and output of the training, it is possible to identify areas of improvement and make necessary adjustments to ensure that the training is effective in meeting the needs of civil service officials and improving their job performance.

D. Phillips ROI Model

The Phillips ROI model evaluates the training program's return on investment (ROI). This model basically emulates the scope and sequence of the Kirkpatrick's Model, but with an additional step. The five levels of the model are as follows:

i. **Level 1: Reaction** - Similar to the Kirkpatrick Model, the model starts with evaluating the participant's reaction and satisfaction to the training program.

ii. **Level 2: Learning** - This level measures the skills and knowledge gained by the participants.

iii. **Level 3: Application and implementation** - Similar to the Kirkpatrick Model, this level measures whether the participants of the training program learned anything from training upon returning to the workplace. However, it improves that scope by determining whether an issue (if any) resides with the application of the learning or its implementation.

iv. **Level 4: Impact** - In the Phillips model, instead of results it focuses on measuring the overall business impact of the training program. It is much broader compared to Kirkpatrick's model as it identifies whether other factors aside from training affected the outcome.

v. **Level 5: Return on investment (ROI)** - This added level is designed to measure the ROI with the use of cost-benefit analysis.

In short above four models steps are;

Will Thalheimer's Training Evaluation Model	*Hamblin-Kirkpatrick Model*	*CIRO Model*	*Phillips ROI Model*
1. No participation 2. Attendance 3. Completion 4. Competency 5. Performance 6. Transfer 7. Business results	1. Reaction 2. Learning 3. Behaviour 4. Result	1. Content 2. Input 3. Reaction 4. Output	1. Reaction 2. Learning 3. Application and implementation 4. Impact 5. Return on investment (ROI)

9.6 Process of Evaluation - Evaluation at the Beginning, During and End of the Program

Process of evaluation: In general, evaluation processes go through four distinct phases: planning, implementation, completion, and reporting. While these mirror common program development steps, it is important to remember that evaluation efforts may not always be linear, depending on where we are in our program or intervention.

A. Evaluation at the Beginning

This involves finding out whether our learners have gained new knowledge and skills. Assess learners' ability to apply new knowledge and skills to their work. Measure payoffs for company as a whole. A payoff can be an improvement in employee performance, a reduction in costs, or increased profits.

The learner's skills and knowledge are assessed at the beginning of the training program and termed, 'entry test'. The purpose of this evaluation is to determine the level of knowledge and skills of the trainees before the training. This will help in making the trainees aware of their state of knowledge and areas where they are lacking, guide the trainers in planning the training program and evaluating the gain made, and sensitize participants to key concepts.

B. Evaluation during the Training

This evaluation is conducted when the instruction starts. It usually consists of short tests at regular intervals. The evaluation during the training is for both the trainer and the participant to know how well they are doing. This helps to improve the learning process and learning principles (*e.g.*, adult learning).

Some of the key principles of this type of evaluation are given below.

This evaluation takes place during classroom lectures, discussions, tests, class problems and demonstrations. Brief examples of these types of evaluation are questions and answers. The purpose of this evaluation is to obtain responses to specific questions.

C. Evaluation at the End of the Training

The evaluation at the end of the training can help to identify:

i) The trainees' reactions to the course and the learning experience.

ii) The learning or changes in attitudes, knowledge and skills of the participants.

iii) The participants' evaluation of the performance of the staff.

This evaluation cannot ascertain the impact on job behaviour. It can only evaluate what was learned. The evaluation forms need not be excessively long. Short forms with specific questions often provide more usable information. This evaluation will broadly cover the aspects, below.

a) **Participants' Reactions:** Participants' reactions to the training programme are to be included in an evaluation form developed for the purpose.

b) **Exit Test:** Where an entry test was given, an exit test can be used to ascertain what learning actually took place. Some training institutions are using the exit test very effectively. These tests are developed around the objectives of the training programme. For example, a sample of test question can be used related to,

   i. Facts participants need to know
   ii. Key concepts and principles
   iii. Techniques, *i.e.*, skills, methods, and procedures
   iv. Processes and where, when, and how to use them
   v. Professional attitudes or values needed by participants.

   The exit test need not be too long. The purpose is to assess behaviour and knowledge.

c) **Sample Interview**: In addition to the final test, some trainers select a 10 per cent random sample of participants for an intensive face-to-face personal interview. This is time consuming but can yield some useful information. Again, this should cover Training Evaluation.

d) **Evaluation of the Trainers by the Participants**: Trainers also need to know how well they have performed. Various evaluation forms can be developed. Without this evaluation, a trainer may not know the areas in which improvement is required.

e) **Staff evaluation of the training programme:** The evaluation of a training programme by the staff of a training organization is followed by some training organizations.

9.7 Statistical Tools for Evaluation of Training

The training coordinator may take the help of a statistician for the evaluation of a training programme. It may be used to see the difference between pre and post test result of the trainees. The statistical analysis may be conducted internally or contracted out to external consultants if the trainees are large in number.

Simple calculations that are most useful and can be easily computed by spreadsheet application include: Frequency, Percentage, Mean, Mode, Median, Range, Variance, and Standard Deviation Sometimes, advanced spread sheets to calculate t-tests, correlations, regressions, *etc.* can also be used.

9.8 Use of Evaluation Findings

Sometimes evaluations done but the results are not adequately utilized. Evaluation should never be done as a formal procedure. The purpose of evaluating a training programme is to determine its value or to identify ways for its improvements. The three major uses of the findings of evaluations are for improving the training programme, improving the institute's training programme and promoting the value of the training.

a) Improving the Quality of the Training Programme

The results of the evaluation should help the training organizers make decision regarding several key questions.

i. Adequacy of time
ii. Meeting objectives
iii. Proper use of methods and aids
iv. Fulfilling needs and interests of participants
v. Adequate field exposure.
vi. Means and ways to improve training program

b) Training Institute's Capability Strengthening

The training institute is interested in improving its programme. Training programme evaluations can provide useful information about how to do this. Some aspects are as follow;

i. Adequacy of facilities
ii. Demand for the programme
iii. Continuation of training program
iv. Improvement in performance level of various trainers

v. Adequacy of budget.
vi. Better scheduling of program
vii. Support staff for the training programme

c) Promotion of Training

Evaluation reports are useful for promoting and sharing information about training programme.

9.9 Indicators of Training Effectiveness

To effectively evaluate training programs, practitioners of the training program must first define the indicators of "effectiveness." Below are some examples of factors or indicators that can help in measuring training effectiveness.

i. New skills and knowledge
ii. Learning experience
iii. Employee happiness
iv. Cultural impact
v. Efficiency impact
vi. Financial impact

Chapter 10

Impact Assessment

This Chapter Contains

Sl.No.	*Content*
10.1	Assessment - Meaning, Need, Features, Benefits, Concepts
10.2	Indicators for Impact Assessment - Direct, Indirect or Proxy, Quantitative and Qualitative Indicators
10.3	Methods of Impact Evaluation
10.4	Learning Retention of Participants (KOSA)
10.5	Impact on the Job Performance
10.6	Impact on Organizational Effectiveness
10.7	Benefits of Organizational Effectiveness
10.8	Impact Assessment on Stakeholders

10.1 Assessment - Meaning, Need, Features, Benefits, Concepts

A. Meaning of Impact Assessment

Impact assessment involves the assessment of long-term and/or significant changes brought about through a development intervention or series of interventions. Impact assessment is not a tool or methodology. Instead, it can be carried out through many different tools, methodologies and approaches.

The significance of an impact is based on expert judgement of the sensitivity) of a receptor and the magnitude (or consequence) of the effect that will be caused by a project-induced change.

B. Need of Impact Assessment

An impact evaluation provides information about the observed changes or 'impacts' produced by an intervention. These observed changes can be positive and negative, intended and unintended, direct and indirect. An impact evaluation must establish the cause of the observed changes.

Five Reasons Why we need an Impact Assessment

1. Indicates direction for other similar projects. It is more so in case of business projects in terms of resource planning and capacity building abilities
2. It helps in planning relevant changes. The impact assessment would provide guideline as how appropriate steps to be taken in planning and executing new projects
3. Drive discussion with stakeholders about future outcomes, consequences, providing insights and better understanding of the projects to be taken up.
4. It helps to determine, scope, scale of changes, their magnitude and impact on society.
5. Environmental Impact Assessment (EIA) will benefit people dealing with biodiversity, flora, ecosystem, water and air in terms of positive and negative impact due to climate change and means of overcoming related problems.

C. Features of Impact Assessment

Identifying, predicting, and evaluating economic, environmental and social impacts of development activities. Providing information on the environmental consequences for decision making. Promoting environmentally sound and suitable development by identifying appropriate alternatives and mitigation measures.

The salient features of Environmental Impact Assessment (EIA) are:

1. Identification of Potential Impacts: EIA involves systematically identifying the potential environmental impacts, both positive and negative, that may arise from a proposed project or activity.
2. Baseline Data Collection: EIA requires the collection of comprehensive baseline data on the existing environmental conditions, including information on the physical, biological, and socio-economic aspects of the project area.
3. Impact Prediction and Evaluation: EIA involves the use of analytical tools and models to predict the magnitude, extent, and significance of the identified environmental impacts, both short-term and long-term.
4. Mitigation Measures: EIA facilitates the development of appropriate mitigation measures to avoid, minimize, or compensate for the identified adverse environmental impacts.
5. Alternatives Analysis: EIA mandates the evaluation of different project alternatives, including the "no-action" alternative, to identify the option that best balances environmental, economic, and social considerations.
6. Public Participation: EIA processes typically involve public consultation and the incorporation of stakeholder inputs to ensure transparency and address community concerns.

D. Benefits of Impact Assessment

Impact assessment provides benefits like ways to reduce negative impacts of the projects, provides information whether proposed project could attain desired goals and selection of alternatives or to modify project to yield positive results.

Key benefits of impact evaluation:

1. Impact evaluation shows the success or failure of a project and holds all stakeholders, including donors and beneficiaries, accountable.
2. By identifying the magnitude of the impact and how it occurred, it aids in determining whether and how effectively an intervention worked to bring about a change in a specific community of interest.
3. Impact evaluations are also helpful in identifying the actual needs on the ground and providing answers to project or program design questions.
4. They help identify which alternative, among several options, is the most effective approach, represents the most significant benefits to the target communities, offers the best value for money, and is the most suitable for scaling up and replicating.
5. It gives business organizations the data to decide whether to alter a current initiative or plan for new actions. Impact evaluation aids organizations in advocating for behavior, attitudes, policy, and legislation changes at all levels using the evaluation's findings.

E. Concepts of Impact Assessment

The term "impact" is conceptually and mathematically defined to be the difference in the state or value of an ecosystem with versus without the source of impact. Some resulting problems associated with the measurement of impacts based on comparisons of baseline and operational data are discussed briefly

10.2 Indicators for Impact Assessment - Direct, Indirect or Proxy, Quantitative and Qualitative Indicators

In general an indicator is a measure of output against input.

A. **Direct indicators** correspond precisely to results at any performance level. It may be in case of training, field activities like irrigation, communication *etc.*

B. **Indirect or "proxy" indicators** are used to demonstrate the change or results where direct measures are not feasible. Some objectives, particularly impact objectives, are difficult to monitor. It is often necessary to select indirect or proxy indicators that may be easier for evaluators to measure. For example behavioural changes like knowledge, skill and attitude

C. **Quantitative indicators** are expressed on a percentage or share basis, or, in terms of rate of change or ratio. For example, reduction in infant mortality and maternal mortality are quantitative indicators of health.

D. **Qualitative indicators** can be defined as people's judgments and perceptions about a subject. For example: the number of people, or, per cent of people, that have adopted farm mechanization in a village after receiving training, is a quantitative indicator.

10.3 Methods of Impact Evaluation

Impact assessment is typically carried out through a systematic and structured process, which includes the following key steps: Scoping: Defining the boundaries and objectives of the assessment. Baseline Data Collection: Gathering data on the existing conditions in the affected areas.

Impact evaluations can be divided into two categories: prospective and retrospective. Prospective evaluations are developed at the same time as the program is being designed and are built into program implementation. Baseline data are collected prior to program implementation for both treatment and comparison groups

What is Impact Evaluation?

Definition

An impact evaluation relies on rigorous methods to determine the changes in outcomes which can be attributed to a specific intervention based on cause-and-effect analysis. Impact evaluations need to account for the counterfactual what would have occurred without the intervention through the use of an experimental or quasi-experimental design using comparison and treatment groups.

Purpose of Impact Evaluation

Impact evaluations often serve an accountability purpose to determine if and how well a program worked. Impact Evaluations can also help answer program design questions to determine which, among several alternatives, is the most effective approach.

When to do an Impact Evaluation?

The World Bank has developed the following guidelines for determining when an impact evaluation may be useful·

i. If it is an innovative intervention scheme, such as a pilot program.
ii. If the intervention is to be scaled up or replicated in a different setting.
iii. If the intervention is strategically relevant and will require a great deal of resources.
iv. If the intervention is untested.
v. If the intervention results will influence key policy decisions.

Design of Impact Evaluations

There are many design options for impact evaluations depending on a variety of factors However, all designs generally fall into either an experimental or quasi-experimental design through the use of a counter factual, an alternate control group generated by random selection in the case of experimental designs, or a comparison group which is not randomized in the case of a quasi-experimental design.

Common Experimental Designs include the following:

i. Randomized Controlled Trial (RCT)
ii. Randomized Offering of Intervention
iii. Randomized Promotion of Intervention
iv. Multiple Treatment Design
v. Common Quasi-Experimental designs include the following:
   a) Difference-in-differences
   b) Matched Comparisons
   c) Regression Discontinuity
   d) Interrupted Time Series

Impact evaluations are not appropriate for every intervention scheme, as such evaluations require certain data sets and financial resources that may not be available. Evaluators usually work in conjunction with project planners and donors when proposing to conduct impact evaluation.

10.4 Learning Retention of Participants (KOSA)

What is Learning Retention?

Learning retention is the process of transferring new information into long-term memory. This means information taken effectively and are recalled it in the future.

Learning retention is a person's ability to transfer new information into their long-term memory so that it is easy for them to recall and put that knowledge to use in the future. In simpler words, learning retention is all about making new knowledge stick for a long time.

Nine Tips to Improve Learning Retention

There are many outside factors that can affect level of learning retention. Managing career and supplemental studies can be a challenge as brain will be dealing with a great deal of information.

Family and home commitments can also add extra pressure and stress, and may not have enough study time or the environment we need. This makes it more important than ever to learn to retain new information effectively.

How to Improve Learning Retention?

1. Break down information into bite-sized chunks
2. Group items into categories
3. Test self
4. Learn from mistakes
5. Active engagement
6. Review and summarize topics
7. Learn by teaching others
8. Get plenty of sleep and exercise

Different Methods for Learning Retention

There are different methods for improving learning retention. Psychologists and educators have attempted to define solutions and strategies to allow learners to successfully retain new information.

Here are examples of two powerful learning methods: the 50/50 learning rule and spaced repetition. The two methods complement each other and can be used simultaneously.

A. The 50/50 Learning Rule

The theory behind '**the 50/50 learning rule**' is to retain the information we should learn for half the time and share what we have learned for the other half.

This means committing 50 per cent of our time to explaining, summarizing, or teaching the information. We'll need to engage with the material enough to explain in a way that others will understand the context. This is why teaching helps to retain information.

In order to put this into practice one could try studying material, such as an online course, summarizing the points and sharing them with a colleague. This will force to think and result in better learning retention.

B. Spaced Repetition

Whereas engaging with the material is crucial for learning retention, repetition is also important. Spaced repetition is based on the idea that if we repeat an intake of the information over a period of time, it is more likely to be remembered.

Our brains learn better when we space out information. It's more effective than trying to retain mass amounts of information together. This allows time for the neural connection to solidify and prevents new information from taking the place of old information.

10.5 Impact on the Job Performance

Effective communication channels, supportive management, and growth opportunities can contribute to higher levels of job satisfaction. The nature of job descriptions and assigned responsibilities can impact job satisfaction.

Naturally, one of the biggest influences on employees' performance are their knowledge, experience, and skills they already possess. Employees without any of these three will most likely perform poorly. This is exactly why it is crucial to offer training to employees who are advancing to more demanding roles.

Top Factors that Affect Work Performance

i. Job Satisfaction.
ii. Employee Engagement.
iii. Training and Development.
iv. The Right Tools for the Job.
v. Company Culture and Work Environment.
vi. Work Environment.
vii. Workload.
viii. Communication.

What are the Factors that Affect Work Performance?

Many factors determine how well an employee can focus, complete tasks, provide good service and be part of the office community. The top factors that affect work performance are,

1. **Job satisfaction:** Sixty per cent of employees consider the people they work with to be "very important to job satisfaction." Here are a few other elements that contribute to job satisfaction.
   i. Leadership is concerned with the needs of their team members
   ii. An employee's knowledge and skillset match their job responsibilities
   iii. Compensation is fair and aligned with industry standards
2. **Employee engagement:** Sixty-eight per cent of employees worldwide feel engaged at work, which, regrettably, leaves 32 per cent unengaged. Engagement is an important factor affecting productivity. Factors that increase employee engagement are,
   i. Employees feel acknowledged and appreciated for their contributions.
   ii. Team members are committed to collaboration within and outside their department.
   iii. There is a sense of community among team members and the company as a whole.
3. **Training and development:** U.S. companies have been increasing their training investments over the last few years. Large companies provide an average of 102.6 hours per employee annually, midsize companies provide 34.7 annual hours, and small companies provide.
4. **The right tools for the Job:** Being provided with the proper tools helps employees perform not only better but faster like,
   i. Office supplies are readily available.
   ii. Physical tools and machines are in good working condition.
   iii. Internet connection is fast and reliable.
   iv. Hardware and software are updated/upgraded as needed.

   Additionally, the importance of technology is very high in performance and productivity. 75 per cent of global organizations are forecast to increase their use of technological productivity tools because they save a lot of time and money.
5. **Company culture and work environment:** When it comes to the factors affecting job performance, few things are as important as company culture. Culture sets procedural and behavioural norms within an organization, including policies, goals, attitudes, and expectations. Organizational culture is a crucial factor for minority employees. As a leader, it can be a bit difficult to decipher the culture and morale within company or department without first taking a step back to observe and contemplate it.

   Here are a few clues that the culture is positive:
   i. No constant complaining.
   ii. Team members seem to have confidence and a sense of purpose.

   iii. Management's conduct is encouraging, not punitive.
   iv. Employees feel safe in and around the property.
   v. The level of turnover is lower.

6. **Work Environment:** As far as the work environment goes, it's often an overlooked factor of work performance. Nevertheless, how the office space looks and feels plays a big role in the comfort and productivity of employees. The following improves performance:
   i. The space and furnishings are clean and tidy.
   ii. The temperature, air quality, lighting, and noise are at comfortable levels.
   iii. Employees can personalize their workspace.
7. **Workload:** Team members who are frequently overwhelmed by a large workload may experience stress, burnout, and decreased productivity. Alternatively, those with little to do may grow disengaged and less productive in their roles. Striking the right workload balance helps individuals stay productive and engaged in their work. In addition, a balanced workload is essential to building a fulfilling career one will genuinely want to invest in for the long term.
8. **Communication:** A business that executes effective communication strategies has a 3.5 times higher chance of outperforming its competitors. Effective communication fosters trust by facilitating prompt and transparent resolution of issues. Additionally, it promotes teamwork and enhances problem-solving. When all individuals have the opportunity to voice their concerns, ideas, and opinions, this nurtures a healthy workplace environment.

Three Methods for Measurement

Three methods work well for gaining the necessary information. Each method should be reviewed before taking any further steps:

1. **Employee feedback:** Asking an employee about how they view their performance can be as effective as it is insightful.
2. **Peer feedback:** Colleagues can speak on their own experiences collaborating and working with fellow employees.
3. **Manager feedback:** Performance reviews and ratings help demonstrate the quality of key performance indicators (KPIs).

10.6 Impact on Organizational Effectiveness

Organizational effectiveness encompasses various factors such as clear vision and goals, strong leadership, strategic planning, efficient resource management, effective communication, talent management, performance measurement, adaptability, and stakeholder engagement.

Factors Impacting Organizational Effectiveness

Organizational effectiveness refers to the extent to which an organization achieves its goals and objectives while utilizing its resources efficiently. It is a measure of how well an organization performs and delivers value to its stakeholders. Key factors that contribute to organizational effectiveness are,

1. **Clear vision, mission, and goals**: An effective organization has a well-defined vision, mission, and goals that provide a sense of direction and purpose. These statements outline the organization's aspirations, values, and the outcomes it aims to achieve. They serve as guiding principles for decision-making, strategy formulation, and resource allocation.
2. **Strong leadership**: Leadership plays a crucial role in driving organizational effectiveness. Effective leaders inspire and motivate employees, provide clear guidance, and make strategic decisions that align with the organization's objectives. They create a positive work culture, foster innovation, and empower employees to perform at their best.
3. **Strategic planning**: A well-developed strategic plan helps organizations align their resources, capabilities, and actions to achieve their goals. It involves analysing the internal and external environment, setting priorities, and formulating strategies to address challenges and leverage opportunities.
4. **Efficient resource management**: Organizational effectiveness relies on the efficient management of resources, including human resources, financial capital, technology, and infrastructure. Effective organizations allocate resources strategically, optimize processes, and continuously monitor and evaluate resource utilization to maximize productivity and minimize waste.
5. **Clear communication**: Effective communication is essential for organizational effectiveness. It ensures that information flows smoothly across all levels and functions of the organization. Clear communication helps align employees' efforts, disseminate goals and objectives, facilitate collaboration, and foster a culture of transparency and accountability.
6. **Talent management**: Organizations that effectively attract, develop, and retain talent tend to be more successful. Effective talent management involves recruiting the right people, providing training and development opportunities, promoting a culture of continuous learning, and recognizing and rewarding performance. By nurturing a skilled and motivated workforce, organizations can enhance their overall effectiveness.
7. **Performance measurement and continuous improvement**: Effective organizations establish performance metrics and regularly measure progress towards their goals. They have mechanisms in place to monitor key performance indicators (KPIs) and evaluate outcomes. This data-driven approach allows them to identify areas for improvement, make

informed decisions, and continuously enhance their processes and performance.

8. **Adaptability and change management**: Organizational effectiveness requires the ability to adapt to changing market conditions, customer needs, and technological advancements. Effective organizations embrace innovation, encourage creativity, and proactively manage change. They are agile and responsive, constantly seeking ways to improve and stay ahead of the competition.
9. **Stakeholder engagement**: Engaging with stakeholders, including customers, employees, suppliers, and the community, is crucial for organizational effectiveness. Effective organizations listen to their stakeholders' needs and expectations, seek feedback, and build strong relationships. By understanding and meeting stakeholder demands, organizations can enhance their reputation, customer loyalty, and overall performance.

10.7 Benefits of Organizational Effectiveness

Organizational effectiveness is important in helping companies flourish long term. It allows organizations to operate more smoothly by keeping them focused on their goals. Benefits of an effective organization include:

i. **Increased productivity:** Setting clear goals and monitoring progress may increase output.
ii. **Strengthened employee engagement:** Employees who know their goals and why they're important may connect more with their work.
iii. **Enhanced management communication:** Taking steps toward organizational effectiveness can help managers align.
iv. **Optimized budget:** Organizational effectiveness typically helps eliminate areas of waste.
v. **Streamlined use of technology:** A company may upgrade or improve its equipment by promoting efficiency.
vi. **Maximized customer value:** More effective organizations may be more capable of meeting customers' needs.

Six Types of Organizational Effectiveness Models

An organization's effectiveness depends on what goal it set out to achieve. Several models help companies determine their effectiveness based on their key objectives:

1. **Internal process model:** In the internal process model, companies focus on internal processes and protocols to make the organization function. By optimizing processes, these companies may improve their operations and increase their productivity. Ways organizations can achieve this include

having a thorough process for documentation and suitable equipment to enable functions.

2. **Goal model focuses:** This is traditional model used to measure an organization's effectiveness. An organization may set various goals, ranging from reaching a financial target to creating a particular societal impact. The goal model might be less actionable than other models, as it focuses on measuring outputs and has little information about the inputs or processes.
3. **Resource-based model:** Organizations using a resource-based model measure their effectiveness by looking at their inputs. This approach focuses on the idea that companies can become successful by obtaining an adequate supply of the resources required. To do this, organizations try to mine rare, valuable and unique resources to create a competitive advantage over other organizations.
4. **Strategic constituency model:** A strategic constituency model is one in which companies evaluate their success by measuring their ability to meet the needs of their environment. Companies that use this approach aim to rise to the standards of parties that control their success. These constituencies can include business owners, employees, customers, suppliers or the government, as each of these groups can influence the organization's survival. To be successful in this approach, businesses identify their constituencies and their expectations so that they can fulfil them.
5. **Abundance model:** An abundance model measures an organization's effectiveness by examining its positive impact on humankind. To companies using this approach, effectiveness means prioritizing appropriate beliefs to maximize human capabilities. Abundance models use a combination of positive and negative values to embrace the full scope of human potential.
6. **Competing values model:** In a competing values model, organizations determine their effectiveness by their ability to create competing values. This approach focuses on different key values like Cameron and Quinn's competing values framework

Tips to Increase Effectiveness of Organization

1. Set goal
2. Make effective decision
3. Develop strategies
4. Consider own needs

10.8 Impact Assessment on Stakeholders

Stakeholder impact refers to the effects an organization's actions, decisions, or policies have on its stakeholders. It's a way of evaluating the impact of an organization's activities on the people, groups, and organizations that are affected by its operations.

By doing so, we can reduce conflicts, enhance collaboration, and increase satisfaction among all parties involved. Stakeholder management can also help achieve project objectives in terms of quality, cost, time, and scope.

What is a Stakeholder Impact Analysis?

A stakeholder impact analysis involves determining how a stakeholder group (or multiple stakeholder groups) may be affected by a project or decision in order to strategically engage with those stakeholders. This process is also known as a stakeholder impact assessment.

Importance of Stakeholder Impact Analysis

i. For include starting a new community development project,
ii. Internally restructuring of organizations
iii. Launching a new product or major updates the product, or
iv. Introducing a new policy or
v. To bring major changes to an existing policy.

How to Conduct a Stakeholder Impact Analysis in 5 Steps?

1. Define the Project	1. Nature of the project 2. Type of change are we are trying to create? 3. Milestones, events, decisions, and steps that may have an impact on stakeholders? 4. When the changes or decisions are going to take place?
2. Identify Stakeholders	There are a number of ways to identify impacted stakeholders 1. Seeing who have been in contact 2. Asking project team members 3. Reviewing contact information 4. Looking on a map for nearby businesses and residents 5. Engaging with the stakeholders already known 6. Considering both directly and indirectly impacted stakeholders

3. Assess Potential Impact	To map state holders we may assign impact level on a five point measurement scale like 1. Very low 2. Low 3. Medium 4. High 5. Very high
4. Analyse Further	Find out criteria like 1. Their level of influence 2. Their interests 3. Their relationships 4. What they know 5. How critical they are to work 6. How they feel about it
5. Plan and Implement	Some important best practices for this step include 1. Check our rights, obligations, and responsibilities towards stakeholders 2. Make sure of addressing the concerns of impacted stakeholders 3. Allow stakeholders to be involved in decisions that may impact them 4. Listen to stakeholders 5. Be open and transparent

Chapter 11

Human Resource Development

This Chapter Contains

Sl.No.	*Contents*
11.1	Human Resource Development: Meaning, Importance and Benefits
11.2	Types of HRD Systems and Sub-systems, Manpower Planning, Recruitment, Career Planning, Succession Planning and Retention
11.3	Work system - Role Analysis, Role Efficacy, Performance Plan, Performance Feedback and Guidance, Performance Appraisal, Promotion, Job Rotation and Reward
11.4	Development System - Induction, Training, Job Enrichment, Self-learning Mechanisms, Potential Appraisal, Succession Development, Counselling, Mentor System, Self-renewal System, Survey and Action Research
11.5	Organisational Development Interventions and Culture System (Vision, Mission and Goals, Values of Organisation, Communication, Get together and Celebrate in HR, Task Force in HR, Small Group in HR)
11.6	Components of HRD System - Performance Appraisal, Potential Appraisal, Task Management System, Development System, Socialisation System in HRM, Governance in HRM, Functions of HRD - Organisational Development, Career Development and Capacity Development

11.1 Human Resource Development: Meaning, Importance and Benefits

A. Meaning of HRD

HRD means a wider range of activities relating to organization and employees. It deals with important aspects like, process of developing and nurturing competency of employees leading to improvement in performance and productivity. The functions of HRD are strategic and holistic approach, fostering organizational culture, continuous learning, improvement of employees, talent acquisition, performance management and career planning, leadership development, and aligns organizational objectives with aspiration of individuals.

B. Importance of HRD

The importance of HRD is as follows.

1. **Enhances employee relations:** HRD enables both employers and employees to maintain good relationship, leading to develop better methods of teamwork, establish proper conduct and set rules for resolving conflicts.
2. **Creates focus on company goals:** Developing an HR strategy can help connect employees more effectively to overall company goals. The HR team may create a series of training videos that educate employees about company vision and goals. These videos may help minimize confusion in team and keep teams better focused.
3. **Clarifies hiring practices:** Developing human resources team can help define hiring practices, leading to greater efficiency and productivity. Better HR practices focus more on company culture and needs could help filter options more efficiently.
4. **Improves workplace operation:** Creating an HR strategy can also improve the operation of workplace. It could help to identify employee engagement and methods of improving it. It could help to track team's interaction, identify obstacles to completion and develop methods of improving operation efficiency.
5. **Increases employee skills:** A human resources strategy may help to improve employees' skills in the workplace.

Levels of Human Resources Development

Most human resource development occurs at three different levels. These elements help focus a company's development and provide a comprehensive HR strategy that helps the business and its employees. These three different developmental levels include:

1. **Individual development**: This element focuses on improving each individual in a company. It strives to teach each employee new information, skills or behaviours that could help them integrate more readily into a workplace and help them grow as people in general.
2. **Career development**: This concept focuses on helping companies find better ways to improve their employees' careers. It may identify methods to increase engagement, add more opportunities for employees and manage their career needs.
3. **Organizational development**: This factor develops creative corporate strategies that improve human resources and overall operation. It could help a team produce a more efficient and effective operation.

C. Benefits of HRD

1. **Improved employee performance:** HRD programs help employees develop the necessary skills and knowledge to perform their jobs effectively. This leads to improved productivity, higher quality of work, and increased efficiency.
2. **Increased employee satisfaction and engagement:** When organizations invest in the development of their employees, it sends a message that they are valued and their growth is important. This leads to higher levels of employee satisfaction and engagement, because employees feel motivated and support their professional development.
3. **Enhanced employee retention:** Organizations that prioritize HRD are more likely to retain their employees. When employees have opportunities for growth and advancement within the company, they are less likely to seek opportunities elsewhere.
4. **Succession planning and talent development:** HRD helps to identify and nurture potential leaders within the organization. By providing opportunities for growth and development, organizations can groom employees for future leadership roles.
5. **Adaptability and agility:** In a rapidly changing business environment, organizations need to be adaptable and agile. HRD equips employees with the skills and knowledge to respond to changing market conditions, technological advancements, and evolving customer needs. This allows organizations to stay competitive and thrive in dynamic environments.
6. **Improved organizational culture:** HRD initiatives contribute to a positive organizational culture. When employees are continuously learning and growing, it fosters a culture of innovation, collaboration, and continuous improvement. This leads to higher employee morale, teamwork, and a shared commitment to the organization's goals.

7. **Knowledge sharing and learning organization:** HRD encourages knowledge sharing and the development of a learning culture within the organization.
8. **Increased organizational effectiveness:** HRD improves the overall effectiveness and performance of the organization. When employees are well-trained, engaged, and empowered, they contribute to the organization's strategic objectives, drive innovation, and achieve sustainable growth.

Objectives of Human Resource Development

The major objectives are,

1. Enhancing Employee satisfaction
2. Promoting Career Growth
3. Fostering Employee Engagement.
4. Building Leadership Capabilities.
5. Encouraging Continuous Learning
6. Improving Performance
7. Supporting Change Management.
8. Enhancing Employee Satisfaction and Retention
9. Promoting Diversity and Inclusion.
10. Aligning HRD with Organizational Strategy

Features of Human Resource Development

Human Resource Development (HRD) is characterized by several key features that distinguish it as a strategic approach to developing and nurturing employees.

1. **Strategic alignment:** HRD is aligned with the strategic goals and objectives of the organization. It focuses on developing human capital in a way that directly contributes to achieving the overall business strategy.
2. **Continuous learning culture:** HRD promotes a culture of continuous learning and development within the organization. It emphasizes the importance of lifelong learning and encourages employees to acquire new knowledge and skills throughout their careers.
3. **Individual and organizational growth:** HRD aims to facilitate the growth and development of both individuals and the organization as a whole. It recognizes that investing in employee development leads to improved performance, engagement, and overall organizational success.
4. **Customization and personalization:** HRD takes into account the unique needs and aspirations of individuals within the organization. It recognizes that different employees may have varying skill gaps and developmental

requirements, and therefore, it offers customized and personalized development opportunities.

5. **Employee empowerment:** HRD empowers employees by providing them with the tools, resources, and support needed to take charge of their own learning and development.
6. **Multi-faceted approach:** HRD encompasses a wide range of development methods and interventions. It may include training programs, coaching, mentoring, job rotations, performance management systems, leadership development initiatives, and more. This multi-faceted approach ensures comprehensive development across various dimensions.
7. **Continuous improvement and evaluation:** HRD is a dynamic process that continuously seeks to improve its effectiveness. It incorporates regular evaluation and feedback mechanisms to assess the impact of development interventions and make necessary adjustments for improvement.
8. **Integration of technology:** HRD leverages technology to enhance the learning experience and make development initiatives more accessible and efficient. It incorporates digital tools, e-learning platforms, virtual reality simulations, and data analytics to support learning and development efforts.
9. **Employee engagement and participation:** HRD actively involves employees in the development process.
10. **Collaboration and partnerships:** HRD often involves collaboration and partnerships with internal stakeholders, external training providers, and industry experts. This collaboration ensures access to diverse perspectives, resources, and expertise for effective development outcomes.

11.2 Types of HRD Systems and Sub-systems, Manpower Planning, Recruitment, Career Planning, Succession Planning and Retention

A. Types of HRD Systems

Human Resource Development (HRD) encompasses various types or approaches that organizations can adopt to develop their employees and enhance their overall capabilities.

Common Types of HRD

1. **Training and development:** This type focuses on providing employees with specific knowledge, skills, and competencies related to their job roles. It includes workshops, seminars, on-the-job training, e-learning programs, and other structured learning interventions.

2. **Career development:** Career development programs aim to support employees in planning and advancing their careers within the organization. This involves providing resources, guidance, and opportunities for skill-building, goal setting, and career progression.
3. **Leadership development:** Leadership development programs are designed to identify and cultivate future leaders within the organization. They focus on developing leadership skills, strategic thinking, decision-making abilities, and other essential qualities for effective leadership.
4. **Performance management:** HRD includes performance management systems that set clear performance expectations, provide feedback, and offer developmental opportunities to enhance employee performance.
5. **Succession planning:** Succession planning aims to identify and develop potential successors for key roles within the organization. It involves assessing readiness for higher-level positions and providing them with targeted development opportunities to prepare them for future leadership roles.
6. **Mentoring and coaching:** Mentoring and coaching programs pair employees with experienced professionals who provide guidance, support, and feedback to facilitate their growth and development.
7. **Talent management:** Talent management encompasses a strategic approach to attracting, developing, and retaining high-potential employees. It involves identifying key talent, creating personalized development plans, and providing opportunities for their advancement within the organization.
8. **Knowledge management:** HRD includes knowledge management initiatives that focus on capturing, sharing, and utilizing organizational knowledge effectively.
9. **Diversity and inclusion development:** This type of HRD focuses on creating a diverse and inclusive workplace culture. It includes initiatives such as diversity training, unconscious bias training, and programs to foster inclusivity, equality, and cultural sensitivity.
10. **Technology-based development:** With the advancement of technology, HRD also includes leveraging digital tools and platforms for learning and development. This can involve e-learning modules, virtual reality simulations, mobile learning apps, and online communities for knowledge sharing.

B. Sub-systems of HRD

1. Performance Appraisal,
2. Potential Appraisal,
3. Career Planning,

4. Training,
5. Organization Development,
6. Rewards,
7. Counseling,
8. Quality Circle,
9. Role Analysis,
10. Human Resource Planning,
11. Procurement and Placement,
12. Job Rotation,
13. Quality of Working Life

1. Performance Appraisal

It is a process to measure performance of an employee. It follows a systematic method. The performance of an employee is measured against the set parameters of knowledge of job, quality and quantity of output, initiative, supervision, leadership qualities, cooperation, dependability, health, versatility, judgment *etc.* On the basis of measurement the rank of an employee is decided. The appraisal also decides the training needs for employees.

The training needs are:

(a) Job analysis.
(b) Establishing standards of performance.
(c) Communicating performance standards to employees.
(d) Measuring actual performance.
(e) Comparing actual performance with standards and discuss with employees.
(f) Initiating corrective actions, if necessary.

2. Potential Appraisal

The potential appraisal measures latent abilities of individuals. It appraises leadership abilities and helps to prepare career planning. The method ensures match between job and abilities of employee. Every organization makes potential appraisal to plan for future.

3. Career Planning

For planning for Career the first and foremost step is to make one's SWOT analysis. He should assess his strength, weaknesses, opportunities and threats.

4. Training

Training is an attempt to improve current or future employee performanc increasing an employee's ability to perform through learning usually by chan the employee's attitude or increasing his or her knowledge. The need for training is determined by the deficiency of employee's performance.

5. Organisation Development

Organization Development is to change beliefs, attitudes, values and structure of organizations that they can better adapt to new technologies, markets and challenges and the dizzying rate of change itself. Organization is an open system and therefore must develop itself by adopting various changes to meet the challenges for constantly changing environment.

6. Rewards

Rewarding employees for their excellent professional performance over and above their usual wages and salaries boost up their morale and encourage others to work hard. Reward systems are to be looked after by HRD. By rewarding hard work employees are motivated further. Output goes up. It leads to maximum utilization of available human resources to enhance productivity. Rewards are one of the ways to remove frustration, depression and dissatisfaction from the employees. Slow work tendencies can also be removed.

Rewards are two types. These are Intrinsic and Extrinsic.

(a) Intrinsic rewards are those that employees receive when they perform well in their job. These are often psychological and can include pride and fulfillment for achieving certain things in the workplace.

(b) Extrinsic rewards: An extrinsic reward is a tangible and visible reward given to an individual or an employee for achieving something. They usually have monetary value such as a salary hike, bonus, award, or public recognition. Description: They are usually offered by the individual's immediate manager who decides upon them.

7. Counselling

Counseling is an essential instrument of HRD. It is a service provided to the employees regarding their personal problems. It reduces employee turnover, absenteeism, tardiness and helps in strengthening the superior subordinate and relationship, understanding problems of juniors and seniors thereby facilitating better communication and effective decision making.

Quality Circle

uality circle is a small group of people doing same or similar work. They meet ily together on regular basis for an hour per week, during office hours eadership of their supervisor and trained to identify, analyze and solve roblems of their work and submit solutions to management.

The objectives of quality circles are summarized below:

(a) It is people building philosophy to develop, enhance and use human resources effectively.
(b) Development of others through the members of quality circle.
(c) Promotion of participative management.
(d) Improvement of supervisory skills of employees.
(e) To promote interpersonal skills and thereby seek resolution of conflict.
(f) Encouragement of creativity and problem solving techniques.
(g) To promote leadership qualities among employees.
(h) To develop quality improvement consciousness among the employees.
(i) To develop group efforts.

9. Role Analysis

Role is a set of expected behavior patterns attributed to a person occupying a given position in a unit. Every employee has a role to play. Everyone does not have to be consistent and regular in one role. Employees have to play diverse roles both on and off the jobs.

10. Human Resource Planning

HR Planning is an integral part of business and organization planning. It concerns with detailed planning to establish the future human resource requirements at every level by skill and category. The quality of human requirements plan depends on the quality of the analysis and data on which it is based.

11. Procurement and Placement

Procurement of human resources is done through recruitment and selection. Vacancies are filled through these two methods. Placement is the process of assigning the procured candidates with suitable jobs. It is placing right persons at right jobs to obtain the best results.

12. Job Rotation

Jobs are rotated among the employees to increase their knowledge and skill required for performance of various jobs. This is essential because during absence or leave of employees work should not suffer.

13. Quality of Working Life

Employees are the assets of an organization. To extract the best from employees adequate, healthy and hygienic working conditions should be provided. This increases the productivity.

C. Manpower Planning

Definition

Manpower planning is evaluating, strategizing, and assessing the workers' skills to create manpower teams that work at optimum productivity within an organization. Improper planning of the workforce can lead to overstaffing or understaffing, resulting in productivity and revenue losses.

Manpower planning - Meaning, types, process, need and factors that affect it:

(i) Meaning

Manpower is the primary resource for any company. In its absence, other resources like raw materials, money, machinery, *etc.*, cannot be used. Workers are needed for every unit, even those with a high degree of automation, such as pharmaceuticals.

Improper planning of the workforce can lead to overstaffing or understaffing, resulting in productivity and revenue losses. Manpower planning, as part of human resource management, empowers managers to create schedules and manage workers while improving production, work culture, and industrial relations.

(ii) Types of Manpower Planning

1. **Manpower planning at an organizational level:** For the overall development of an organization, manpower planning can be done at an organizational level to create more opportunities for employment while utilizing available human resources most efficiently.
2. **Manpower planning at micro-level:** A more practical approach to manpower planning is at the micro level. At this level, managers take various measures at the business unit's recruitment, selection, promotion, and expansion stages to train, compensate and plan workers.
3. **Short-term manpower planning:** When manpower planning is done for one year, it is called short-term manpower planning. Annual plans are beneficial at the company level because short-term gains can be integrated easily into the company's positive medium-term or long-term growth.
4. **Medium-term manpower planning:** Generally, manpower planning that is for a period of 2 to 5 years is considered medium-term. Training and development of employees is often part of medium-term Manpower planning as it focuses on imparting skills that will be useful for the company in the next five years.
5. **Long-term manpower planning:** Any manpower planning scheme that is over ten years is long-term manpower planning. Most companies utilize long-term manpower planning to forecast the growth of the company and plan further expansions.

(iii) Process of Manpower Planning

Before managers can forecast future manpower planning, they must evaluate the current manpower inventory. The type of manpower planning in HRM that will be suitable for company depends on the kind of organization, the number of departments, and the employees in each work unit.

Manpower planning is multidisciplinary and utilizes several forecast models. It usually includes.

Model	*Description*
1. Trend Analysis	Managers have to use extrapolation (projecting past trends), indexation (using the current year as the basis), and statistical analysis to anticipate any sudden demand or decline in the number of workers in the company.
2. Work Load Analysis	It involves evaluating the nature of workload in a department, a branch, or the division to map it to the available manpower.
3. Developing employment programs	Once accurate future forecasts are made based on the current manpower inventory, one can create employment programs to hire or train more the number of workers as per the forecast.
4. Design training programs	Training and development programs can make workers more productive at their jobs. The training programs depend on the extent of improvement required in the workforce.

(iv) Need for Manpower Planning

The need for manpower planning arises from the following requirements of a company:

1. In case the shortage of manpower or in future.
2. To decide recruitment of hiring of man power.
3. To reduce labour costs from hiring excess staff or overstaffing
4. Manpower planning can also help identify the available talent.
5. It also supports the growth and diversification of business.

(v) Factors that Affect Manpower Planning in HRM

Manpower planning requires coordinated efforts from managers, supervisors, and executives. The following factors must be considered for manpower planning in HRM.

1. **Monitoring of current manpower:** Available manpower must be closely monitored to evaluate their productivity and efficiency. Automation software like Truein can significantly simplify the process of staff time and attendance management.

2. **Participation of all stakeholders:** Management, top-level executives, and workers must be part of manpower planning. Low-level workers can provide important information regarding the work environment, resource utilization, and training requirements.
3. **Socio-economic work culture:** Manpower planning in HRM is also affected by the work environment within an organization. If the workers are unsatisfied, continuously stressed due to work, or have an unhealthy work/life balance, it can adversely affect manpower planning efforts.

D. Recruitment

Definition

Recruitment is the process of actively seeking out, finding and hiring candidates for a specific position or job. The recruitment definition includes the entire hiring process, from inception to the individual recruit's integration into the company.

Recruitment is the process of finding, selecting, and hiring the best-qualified candidates for available positions within an organization in a timely manner. Potential candidates may be sourced externally from outside of an organization or internally from a group of existing employees who possess the required skills for the position and desire to fill an open role.

The concept of human resources relates to both the people who work for an organization as well as the department within a business that is responsible for the management of all matters pertaining to employees. The human resources department is a vital component within any large company because they are tasked with functions such as staffing, training, driving organizational development, and maintaining human relations. Recruitment is another important role associated with human resources and it is used heavily in most fields of business.

Purpose of Recruitment

The main goal of recruitment is to enable an organization to hire the well-suited candidates from the job market to fill unique and specific roles within their business. Good recruitment means that when the best talents are hired, an organization is able to remain competitive and specialized in their respective fields.

Stages of Recruitment Process

The recruitment function of HR includes analysing the requirements of an available job, attracting suitable candidates to the position, carefully screening and selecting all applicants, hiring, and seamlessly integrating new employees and roles into the organization. There are four stages of the recruitment process used by human resources departments to successfully and comprehensively achieve each of these objectives:

i. Job analysis
ii. Advertising

iii. Screening of applicants

iv. Finalization

i. Job Analysis

The initial step of the recruitment process is job analysis. During this stage of recruitment, the human resources department determines the need for additional personnel and begins to review and analyze what skills may be required of prospective employees in the available position. Based on the analysis, human resource divisions are able to create a detailed job description.

ii. Advertising

After a job description has been created, the human resources department begins to disseminate information about the available job through advertising and search for applicants. Social networking, commercial advertising, and other forms of creative recruitment are effective methods of advertising that have become increasingly popular in recent years.

iii. Screening of Applicants

Screen for candidates, we need to complete three steps: Review resumes and cover letters. Conduct a video or phone interview and identify top candidates.

What is Candidate Screening?

Definition: Candidate screening is the process of reviewing job applications. This comes in right after candidate sourcing and involves skimming through resumes and cover letters to find the closest applicant-job-description matches keeping in mind qualifications, experience, skill sets, and projected candidate-organization fit.

A three step guide to the candidate screening process;

Step 1: Ticking off the basic or must-have requirements

Step 2: Scanning for preferred or good-to-have qualifications

Step 3: Matching the holistic picture of the candidate to the role

iv. Finalization

Typical steps in the recruitment process include the following:

a. Conduct a job analysis to determine position requirements.

b. Develop a job description to include skills and competencies, experience and education, as well as position-specific requirements.

c. Evaluate the job by comparing it to other positions in the organization to determine the overall value of the position.

d. Get the job description and pay range approved.

e. Develop a job posting

f. Use an applicant tracking system **(ATS)** to post the job.

g. Educate everyone involved in the hiring process to ensure awareness of the laws and company guidelines they need to follow.

h. Use recruiting software to screen submitted resumes.

i. Perform a first interview, possibly by phone, to confirm basic information and gauge the candidate's interest.

j. Develop and share interview process guidance to help hiring managers conduct good interviews.

k. Conduct a second interview with the hiring manager or hiring team.

l. Send a pre-hire assessment to the candidate. This can be a personality type test, a coding test for software development roles or similar assessments.

m. Perform reference checks with current and former employers and personal references.

n. Carry out background checks, such as validating the candidate's education, and does a criminal history check.

o. Have the candidate take a drug test, if that's legal in your jurisdiction and required for the position.

p. Negotiate the contract, including salary or hourly rate, number of vacation days, bonus eligibility and other items the company offers or the candidate requests.

q. Send the job offer to the candidate for review and acceptance.

r. Hire the candidate, and move them to the on boarding process.

E. Career Planning

Definition

Career development is a process that helps individuals plan careers that match the company's requirements and help them meet their objectives. It is an ongoing process for each employee where the organisation's duty is to share training and development programmes that may help employees progress along their career paths.

The employee usually requires assessing their skills and experience to better understand their strengths and areas to improve to become more proficient as a professional. They may then discuss with their supervisors about the opportunities in those areas or fields.

Benefits of Career Planning

There are various benefits that come with planning of career

1. Increases employee motivation
2. Betters employee commitment

3. Increase morale
4. Improves organisational support
5. Helps in identifying employee skills and interests
6. Retaining employees
7. Reduces recruitment cost

F. Succession Planning

Definition

Succession planning is the strategic process of identifying and developing internal candidates to fill key organizational roles to ensure business continuity and leadership in the face of role transitions, departures, or unforeseen events. It is relevant to all companies, large or small, for-profit or non-profits.

The Process of Succession Planning - 5 Steps to Consider

i. Identify key positions in practice.
ii. Establish clear criteria for succession.
iii. Assess current talent.
iv. Identify potential successors and provide mentorship and coaching.
v. Develop and document of succession plan.

Different Types of Succession Planning

1. Internal succession planning.
2. External succession planning.
3. Leadership succession planning.
4. Non-leadership succession planning.
5. Long-term succession planning.
6. Emergency succession planning.
7. Transition succession planning.
8. Ensure it is a long-term plan.

Benefits of Succession Planning

1. Decrease hiring expenses
2. Reduce time to hire and time to fill
3. Support continuous workflow despite disruptions
4. Promote engagement and retention
5. Have higher performance
6. Plan and prepare for retirements.

Succession Planning and Diversity

One significant advantage of succession planning is its contribution toward helping organizations meets their diversity and equity.

The succession planning allows organizations to strategically develop people to prepare them to take on more responsibility and higher positions, it can be used to remove barriers to internal advancement that may exist because of implicit bias. By making an effort to plan for the promotion of candidates from diverse backgrounds, we can ensure to maintain equity throughout all levels of organization.

How does the Succession Planning Process Work?

Succession planning may look different from organization to organization, but the general process will typically require the following steps:

1. Identify critical and vulnerable positions in need of successors.
2. Develop a profile of each position and its performance expectation to determine eligibility requirements.
3. Determine which positions are well-suited to transition into the successor position if necessary.
4. Nominate successors from the qualified positions.
5. Create a plan for developing potential successors' skills, abilities, and knowledge to prepare them for the role.
6. Manage individual talent development plans for all high-potential succession candidates, ensuring they are preparing for their potential new role.

Best Practices for Succession Planning

HR leaders must ensure that the succession plan satisfies company needs and people's career objectives. The following best practices can help to create an effective succession plan that promotes employee-employer alignment:

1. **Design a succession planning strategy:** The strategy should outline objectives and methods HR leaders will use to determine which jobs to train for, the individuals to include in the pipeline, the training materials and resources, different educational approaches to use, the desired outcome of the program, the budget, and the timeline.
2. **Begin with a trial program:** Instead of jumping into the water unprepared, HR leaders can get their feet wet by initiating an experimental succession plan.
3. **Evaluate people's competencies:** By assessing people's skills and knowledge, HR leaders can make educated decisions regarding filling positions.

4. **Implement a learning and development program:** A learning and development program nurtures a growth-oriented culture. Integrating various learning methods into the work culture gives people the training they need to assume new positions.
5. **Assess the results.** HR leaders should use key succession planning metrics to analyse the program's effectiveness following its completion.

What are the Risks of Neglecting Succession Planning?

Not only does succession planning have major benefits for organization, but there are also risks involved in neglecting it.

For example:

i. Loss of essential talent and knowledge
ii. Increase in recruitment and training costs
iii. Uncertainty and confusion for employees
iv. The business can lose sight of its mission
v. It harms the company's ability to adapt

Succession Planning Model

The succession planning model depicted below is an adaptation of Groves (2005), who proposed an integrated leadership development and succession planning model. Talent is developed, and leadership talent is spotted, identified, actively developed in leadership programs, and prepared for succession.

Preconditions for Succession Planning

Our succession planning framework has two preconditions:

(i) Creation of a culture of leadership development
(ii) Selection of critical roles

Precondition 1: A Culture of Leadership

For succession planning and long-term leadership development to succeed, there must be a clear commitment from senior management. In addition, managerial performance appraisal and reward processes should incorporate leadership development activities, including the identification and development of high-potential employees.

Precondition 2: Selection of Critical Roles

The easiest way to identify these roles is to look at the most senior people in the organization's hierarchy or the highest earners. If the organization has a well-defined job architecture with function descriptions and rewards based on responsibilities, this selection will be fairly accurate.

A Four-Step Succession Planning Process

Once the two preconditions are met, we can put succession planning process in motion. The four-step process is based on the succession planning model discussed above. The key process steps are,

1. Talent development
2. Identification of leadership talent
3. Leadership development
4. Succession decision

1. **Talent development:** Taking a long-term approach to succession planning and combining it with talent and leadership development is two of the best practices when creating a succession planning strategy. Because succession planning generally focuses on senior leadership roles that would be replaced by employees currently in mid-level leadership roles.
2. **Identification of leadership talent:** During talent development programs, not everyone will succeed, which is why we need to continuously monitor and evaluate the participants. Those who aren't suitable to fill critical leadership positions may be great candidates for low or middle-management positions.
3. **Leadership development:** In this stage, we develop those identified as having strong leadership potential further. This differs from our average talent development. For example, we organize internal courses and workshops for future leaders.
4. **Succession decision:** Based on the previous steps, we can create a succession matrix. In this matrix, we map potential successors for each function.

G. Retention

Definition

Employee retention is the organizational goal of keeping productive and talented workers and reducing turnover by fostering a positive work atmosphere to promote engagement. This includes showing appreciation to employees, providing competitive pay and benefits, and encouraging a healthy work-life balance.

Importance of Employee Retention

Employee retention is important for team building and cohesion in the workplace so workers can trust and depend on each other.

Diminished productivity and competitive advantage are among the biggest losses when talented employees leave an organization. High employee turnover rates can harm an organization's ability to carry out its mission due to impairments

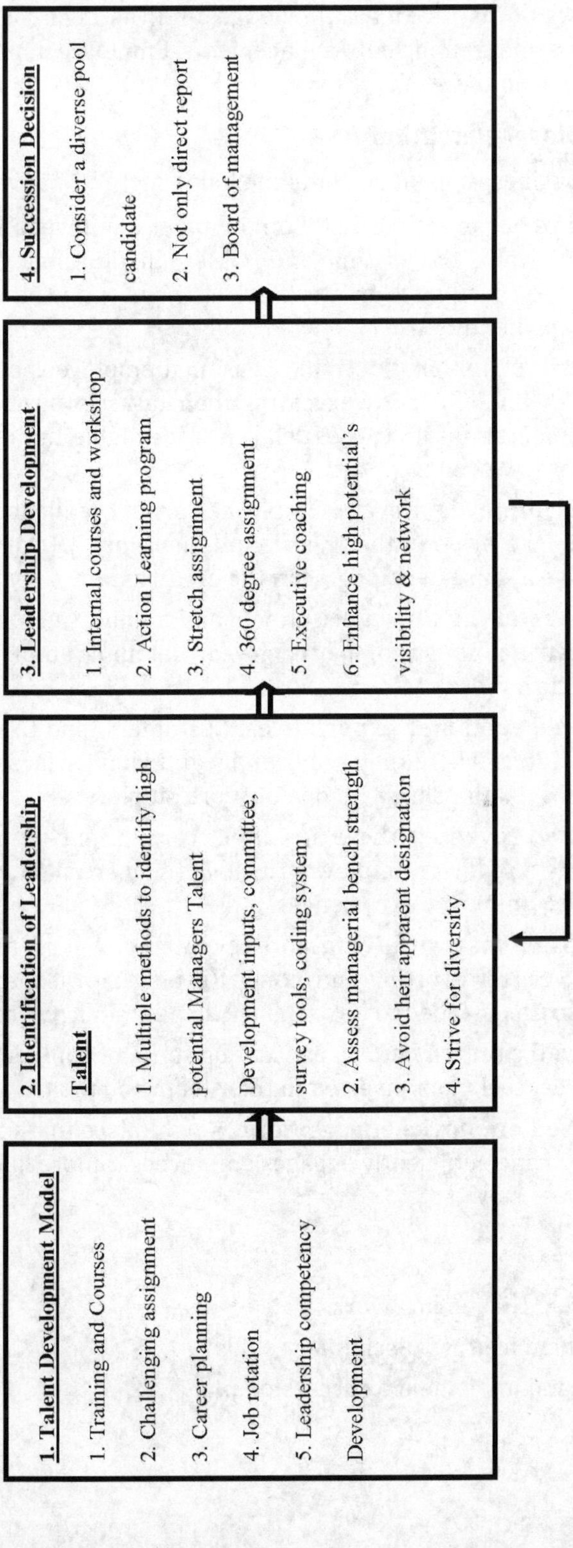
1. Talent Development Model
1. Training and Courses
2. Challenging assignment
3. Career planning
4. Job rotation
5. Leadership competency Development
2. Identification of Leadership Talent
1. Multiple methods to identify high potential Managers Talent Development inputs, committee, survey tools, coding system
2. Assess managerial bench strength
3. Avoid heir apparent designation
4. Strive for diversity
3. Leadership Development
1. Internal courses and workshop
2. Action Learning program
3. Stretch assignment
4. 360 degree assignment
5. Executive coaching
6. Enhance high potential's visibility & network
4. Succession Decision
1. Consider a diverse pool candidate
2. Not only direct report
3. Board of management

to continuity, loss of institutional knowledge, and the high costs of replacing departing workers and training new replacements. Employee departures can also lower the morale of employees.

Benefits of Employee Retention

Significant business benefits include the following:

1. **Better process efficiency.** Longer-term employees know how the company wants things done and possess institutional knowledge they can draw on, leading to greater work efficiencies and attaining company goals expediently.
2. **Greater worker productivity.** Seasoned employees are often skilled at quickly and efficiently executing their tasks. Conversely, hiring new replacements typically causes delays and inevitable, prolonged and costly mistakes in workflows.
3. **Higher employee morale.** Employees are more likely to have higher morale, take pride in their work and, therefore, perform better when they have a sense of belonging in the organization.
4. **Reduced staffing costs.** Recruitment and training can cost organizations significant amounts of money each year. Retaining employees is the best way to curb those costs.
5. **Improved customer experiences.** Customers tend to strongly favour organizations where they see friendly and familiar faces over time and develop a relationship with one or more employees
6. **Increased revenue.** There are direct correlations between increased revenues and lower employee turnover, higher employee morale and improved employee experiences.
7. **Decreased costs of hiring and recruiting.** A high turnover rate is expensive in terms of time and money. Keeping knowledgeable employee's saves on time needed to hire, on board and train new employees.
8. **Improved company culture.** A strong sense of company culture helps employees feel more invested in their team and their work.
9. **Increased employee engagement.** A workplace that values employee retention and sufficiently engages employees is more likely to see those employees stay.

Strategies

1. Give them challenging work
2. Train them in cross-disciplinary skills
3. Define and implement a succession plan

Strategies for Average and Poor Performers

Average Performers	*Poor Performers*
1. Offer personalized benefits and perks	1. Identify the cause of poor performance
2. Ensure they are working under the right manager	2. Address skill gaps immediately
3. Adopt a social recognition system to recognize them	3. Work system analysis

11.3 Work System - Role Analysis, Role Efficacy, Performance Plan, Performance Feedback and Guidance, Performance Appraisal, Promotion, Job Rotation and Reward:

A. Role Analysis

Definition

The Work System Method (WSM) is a system analysis method for thinking about systems in organizations whether or not complete data and technical expertise are available. WSM's goal is to help business professionals understand and communicate about a system at whatever level of detail is appropriate.

Role analysis is a systematic examination of the purpose, responsibilities and scope of a role in supporting business objectives.

It is the foundation for informing the role description and establishing the standards and other essential requirements needed to perform the role.

A thorough role analysis will support to:

i. Produce a high-quality role description that gives clear information to potential candidates, assessors and managers and informs how the role is graded
ii. Attract suitable candidates and select the best person for the role
iii. Consider ways to structure the role (how, where and when work is performed) to enable flexible working and workplace adjustments
iv. Develop appropriate, non-discriminatory pre-screening and disqualification questions
v. Design capability-based assessments around the pre-established standards for the role.

When to Do a Role Analysis?

The initial role analysis is done when the need for a new role is identified. In many cases there may be an existing role description that can be adapted for the new role.

Recommended Approach

An effective role analysis draws on a range of information such as:

i. The hiring manager's description of their needs in the context of the needs of the team and organisation
ii. Interviews with stakeholders who understand the purpose and outcomes of the role (such as supervisors, customers, persons currently in the role or similar role)
iii. Review of an existing role description or similar role description in the sector role description library
iv. Research and comparisons with roles in other agencies, clusters, or external organisations.

A multi-method approach is recommended to give a fair picture of the role and the required capabilities. Well established role analysis techniques include:

i. Critical incident interviews that seek information on behaviours that differentiate between success and failure
ii. Visionary interviews forward looking interviews with leaders to anticipate future directions
iii. Structured role analysis questionnaires - questionnaires to add rigour to the job analysis
iv. Review of existing role-related documentation
v. Direct observation of a person performing the role (useful for roles involving manual tasks).

Outputs of the Role Analysis

The main output of a role analysis for recruitment is the role description. Guidance on what to include in role descriptions is provided in the Role Description Development Guideline.

Tips for Doing a Role Analysis

The role analysis should:

i. Be based on regular reviews and workforce planning projections of needs
ii. Focus on the business outcomes to be delivered by the role and the ways in which the capabilities, experience, knowledge and other essential requirements support this
iii. Consider any essential requirements needed to perform the role from day one
iv. Be objective and be based on a range of evidence
v. Rethink the where, when and how work can be done. Also consider who can do it. Consider modern ways of working, flexibility and allowing for adjustments to accommodate people with disability

vi. Look beyond the current role structure and consider options for reshaping the role to open it up to other candidates

vii. Use or adapt existing role descriptions when the new role is similar to an existing role or role type to maintain consistency across agency's role descriptions

viii. Be centred on the requirements of the role and how they help to meet business outcomes to develop a high quality role description and fit-for-purpose assessment approach.

B. Role Efficacy

Role efficacy is dependent on own potentiality technical competency, managerial skill, experience and design of role that one has to perform.

Definition

The performance of people working in an organisation depends on their own potential effectiveness, their technical competence, their managerial skills, and experience, and the design of the roles they perform in the organisation. It is the integration of individuals and their roles that ensures their effectiveness in the Organisation. There are two aspects, whether an employee is unable to play role properly or the organization is not providing required facilities to play role effectively. These two aspects are valid consideration in role efficiency.

Dimensions of Role Efficacy

Role efficacy has ten dimensions.

1. **Centrality vs. perpherality:** The dimension of centrality measures the role occupants' perception of the significance of his or her role. The more central that people feel their roles are in the organization the higher will be their role efficacy.
2. **Integration vs. distance:** Integration between the self and the role contributes to role efficacy, and self-role distance diminishes efficacy.
3. **Proactively vs. reactivity:** When a role occupant takes initiative and does something independently, that person is exhibiting proactive behaviour; On the other hand, if he or she merely responds to what others expect, the behaviour is reactive.
4. **Creativity vs. routinism:** When role occupants perceive that-they do something new or unique in their roles, their efficacy is high. The perception that they do only routine tasks lowers role efficacy.
5. **Linkage vs. isolation:** Inter role linkage contributes to role efficacy. If role occupants perceive interdependence with others their efficacy will be high. Isolation of the role reduces efficacy.

6. **Helping vs. hostility:** One important aspect of efficacy is the individual's perception that he or she gives and receives help. A perception of hostility decreases efficacy.
7. **Super ordination vs. deprivation:** One important aspect of efficacy is the perception that the role occupant contributes to some larger entity.
8. **Influence vs. powerlessness:** Role occupant's feeling that they are able to exercise influence in their roles increases their role efficacy. The influence may be in terms of decision making, implementation, advice or problem solving.
9. **Growth vs. stagnation:** When a role occupant has opportunities and perceives them as such - to develop in his or her role through, learning new things, role efficacy is likely to be high. Similarly, if the individual perceives his role is lacking in opportunities for growth, his role efficacy will be low.
10. **Confrontation vs. avoidance:** When problems arise they can be confronted. And attempts made to find solutions contribute to efficacy, and avoidance reduces efficacy.

C. Performance Plan

Definition

Performance planning is a strategic process that ensures that an organization's goals are met by its employees. It involves setting specific, measurable, achievable, relevant, and time-bound goals for employees, and then creating a plan to help them meet those goals. Performance plans offer employers a way of helping employees who are struggling in their position. Instead of directly reprimanding or even terminating employees, employers can propose a helpful plan and offer supporting materials for further aid. Performance plans have specific requirements, however, for more effective outcomes.

What is a Performance Plan?

A performance plan or Performance Improvement Plan (PIP) is a formal business document explaining goals for a particular employee. It also outlines potential performance issues hindering the employee's progress in reaching those goals.

How to Create a Performance Plan?

Performance planning is a structured approach consisting of several steps to achieve outlined goals either for an individual team member or a team within the organization. But before we formulate a plan, it's essential to understand why performance planning is needed.

1. **Automate the process:** Gathering all the data from employees and managers, this information should be stored in a centralized system like

an Intranet or even better, a learning management system and by using a system like an LMS, we can assign these questionnaires to both managers and employees as well as give deadlines for filling everything out.

2. **Explain performance reviews:** While regular sit-down reviews should be part of performance planning, they should not be a formality nor take the place of ongoing communication. It's important during the on boarding process for employees to understand company offers open communication that is open-ended.
3. **Align goals:** Next, look at the company goals and compare these with the goals of the department as well as those of the employees. If everyone did their performance planning properly, this step should be easy.
4. **Define tactics:** After the goals have been set, how will employees meet these goals? What resources will they need to do their jobs properly? Again, this is a great opportunity to utilize a learning management system.
5. **Connect employees to the bigger picture:** No matter what size the company is, employees and even teams can feel disconnected from one another. While it's vital to focus on individual and department goals, it's necessary to keep the big picture in mind.
6. **Discuss performance:** Once the tone is set for the meeting, it's time to tie each employee to the set training objectives. Make sure to start this conversation on a positive note. Point out what each employee excels at and how their job is directly tied to current company success.
7. **Create an ongoing communication plan:** After discussing any issues throughout the **performance process**, create a communication plan for the future. While we've already established open communication lines, this takes the process one step deeper.
8. **Set regular performance reviews:** Regular performance review keeps everybody alert to work towards goal of the organization.
9. **Analyze the performance process:** Performance planning can always be improved. Take the feedback from each employee and create a post-write-up for each review. Then before the next review period, go back and analyze this feedback.

Objectives of Performance Planning

1. Clearly defined business goals
2. Set expectations
3. Open communication
4. Performance standards
5. Determine Training needed

Writing of Performance Plan

Here are some steps to follow when writing a performance plan:

1. Identify performance or behavioural issues that need improvement.
2. Explain the level of performance and behaviour expected.
3. Provide helpful suggestions for employee improvement.
4. Schedule progress meetings with the employee.
5. Outline consequences for not meeting standards.

The following template outlines the necessary sections for a performance improvement plan:

1. Supervisor: [Your name, job title]
2. Team member: [Team member's name, job title] [Date]
3. Reasons: [Reason for PIP issuance]
4. Expectations: [Explanation of required expectations]
5. Suggestions: [Helpful suggestions]
6. Schedule: [Meeting schedule]
7. Consequences: [Potential consequences]
8. Manager's signature: [Signature of manager]
9. Team member's signature: [Signature of team member]

D. Performance Feedback and Guidance

Guidelines

Be clear, thoughtful, and respectful. Make sure the feedback offered is accurate, specific, and is limited only to the behaviour in question. Do not evade or soften criticism that needs to be delivered. Provide a detailed explanation of what is wrong and recommend ways to make adjustments.

Concept of Effective Employee Feedback

Employee feedback is an incredibly powerful tool. If offered properly, it has the ability to grow and develop the people of organization, improve the levels of trust and communication, and strengthen bonds between employees and managers.

Nine Tips to Obtain Effective Feedbacks

1. Avoid giving unsolicited advice
2. Be specific
3. Come with a deep level of empathy
4. Don't wait for a quarterly review
5. Keep it private
6. Don't take the "sandwich approach"

7. Make the conversation a two-way street
8. Focus on performance, not personality
9. Keep the conversation going by following-up

E. Performance Appraisal

Concept

The term "Performance Appraisal" refers to the regular review of an employee's job performance and overall contribution to a company. Also known as an annual review, employee appraisal, performance review, or evaluation, a performance appraisal evaluates an employee's skills, achievements, and growth, or lack thereof.

A performance appraisal is a systematic and periodic process of measuring an individual's work performance against the established requirements of the job. It's a subjective evaluation of the employee's strengths and weaknesses, relative worth to the organization, and future development potential.

Performance appraisals are also called performance evaluations, performance reviews, development discussions, or employee appraisals.

Five steps in the performance appraisal process:

i. Develop an evaluation form.
ii. Identify performance measures.
iii. Set guidelines for feedback.
iv. Create disciplinary and termination procedures.
v. Set an evaluation schedule.

Types of Performance Appraisals

Performance appraisals can be broken down into four distinct significant types:

Types	*Description*
1. 360-Degree Appraisal	The manager gathers information on the employee's performance, typically by questionnaire, from supervisors, co-workers, group members, and self-assessment.
2. Negotiated Appraisal	This type of appraisal uses a mediator to help evaluate the employee's performance, with a greater emphasis on the better parts of the employee's performance.
3. Peer Assessment	The team members, workgroup, and co-workers are responsible for rating the employee's performance.
4. Self-Assessment	The employees rate themselves in categories such as work behaviour, attitude, and job performance.

How Performance Appraisal Work?

Human resources (HR) departments typically create performance appraisals as a tool for employees to advance in their careers. They give people feedback on how well they are doing in their jobs, ensuring that they are managing and achieving the goals set for them and assisting them if they fall short.

Performance evaluations assist in determining how to distribute a company's limited budget for giving out incentives, such as raises and bonuses. In addition, they give businesses a tool to identify the workers who have made the most contributions to their expansion so that they may appropriately reward their top performers.

Methods of Performance Appraisal

Performance appraisals come in many forms. Managers and human resources staff responsible for these appraisals need to choose the best methods based on the size of their organization and what sorts of responsibilities the employees fulfill.

Methods	*Brief Description*
1. 720-Degree Feedback	The 720-degree feedback method collects information not only from within the organization but also from the outside, from customers, investors, suppliers, and other financial-related groups.
2. The Assessment Centre Method	This method consists of exercises conducted at the company's designated assessment centre, including computer simulations, discussions, role-playing, and other methods. Employees are evaluated based on communication skills, confidence, emotional intelligence, mental alertness, and administrative abilities.
3. Behaviourally Anchored Rating Scale (BARS)	This appraisal measures the employee's performance by comparing against specific established behaviour examples.
4. Checklist Method	This method consists of a checklist with a series of questions that have yes/no answers for different traits.
5. Customer/Client Reviews	The method fits best for employees who offer goods and services to customers. The manager asks clients and customers for feedback, especially how they perceive the employee and, by extension, the business
6. Critical Incidents Method	Critical incidents could be good or bad. In either case, the supervisor takes the employee's critical behaviour into account.
7. Field Review Method	The HR representative conducts the employee's performance evaluation.
8. Forced Choice Method	This method is usually a series of prepared True/False questions.

Methods	*Brief Description*
9. General Performance Appraisal	The method involves continuous interaction between the manager and the employee, including setting goals and seeing how they are met.
10. Human Resource Accounting Method	Alternately called the "accounting method" or "cost accounting method," this method looks at the monetary value the employee brings to the company.
11. Management by Objective (MBO)	This process involves finding out progress towards reaching of goal.
12. Performance Tests andObservations	This method consists of an oral test that measures employees' skills and knowledge in their respective fields.
13. Project Evaluation Review	This method involves appraising team members at the end of every project, not the end of the business year.
14. Rating Scales	These ratings measure dependability, initiative, attitude, *etc.*, ranging from Excellent to Poor or some similar scale.

F. Promotion

Promotion refers to the advancement or upward movement of an employee within an organization to a higher job position or rank that carries greater responsibility, authority, and potentially higher salary or benefit.

It is a recognition and reward for an employee's performance, skills, experience, and potential for growth.

Features of Promotion

1. **Advancement in job position:** Promotion involves a change in an employee's job position, typically to a higher-level role within the organizational hierarchy. This could include moving from an entry-level position to a managerial position or from a junior role to a senior role.
2. **Increased responsibilities and authority:** With promotion, employees are entrusted with additional responsibilities, decision-making authority, and the expectation to take on more challenging tasks.
3. **Recognition of performance and potential:** Promotion is often based on an employee's demonstrated performance, achievements, and potential for growth within the organization. It signifies recognition of their contributions, skills, and dedication to their work.
4. **Career progression and development:** Promotion offers employees the opportunity to progress in their careers and develop their professional skills. It provides a pathway for growth, advancement, and the acquisition of new knowledge and competencies.

5. **Salary and benefits:** Promotions are often accompanied by an increase in salary or compensation, reflecting the higher level of responsibility and authority.
6. **Internal opportunities:** Promotions can occur within an organization, where existing employees are promoted to fill higher-level
7. **Performance evaluation and criteria**: Promotion decisions are typically based on a comprehensive evaluation of an employee's performance, skills, potential, and alignment with the organization's strategic objectives. Criteria for promotion may include factors such as job performance, demonstrated competencies, experience, tenure, and organizational needs. Promotion is an employee career growth and organizational success.

Importance of Promoting the Employees

1. Promoting employees is crucial for several reasons
2. Recognition of performance
3. Career development
4. Retention of talent
5. Succession planning
6. Increased productivity
7. Boost to team morale
8. Enhanced employer brand

Overall, promoting employees is not just about filling higher positions but also about nurturing talent, fostering loyalty, and building a strong organizational culture cantered around growth and development.

Types of Job Promotions

There are several types of job promotions that organizations commonly use to recognize and reward employees' contributions and advancement in their careers:

1. **Vertical promotion:** This form of promotion entails advancing an employee within the same job function or departmental structure, moving them up.
2. **Horizontal promotion:** A horizontal promotion involves moving an employee to a different role or position at the same level of hierarchy.
3. **Promotion with added responsibilities:** In this type of promotion, employees are given additional responsibilities and tasks while remaining in their current role.
4. **Promotion to a leadership position:** This type of promotion involves moving employees into leadership roles such as Supervisors, Managers, Directors, or Vice Presidents.

5. **Promotion based on skills development:** Employees may be promoted based on the development of new skills or expertise that are valuable to the organization.
6. **Promotion due to organizational restructuring:** Organizational changes, such as mergers, acquisitions, or reorganizations, can lead to promotions as roles and responsibilities evolve. Employees may be promoted to align with new business structures or strategic objectives.
7. **Promotion as a reward for performance:** Promotions can also be used as a reward for exceptional performance, outstanding achievements, or significant contributions to the organization. This type of promotion motivates employees to excel and demonstrates the organization's recognition of their efforts.
8. **Promotion as succession planning:** Succession planning involves promoting employees to prepare them for future leadership roles within the organization. This type of promotion focuses on developing a pipeline of talented individuals for key positions.

Each type of job promotion serves different purposes and can be tailored to meet the specific needs of the organization and its employees' career aspirations.

How to Promote Employees?

Employee promotion is a strategic process aimed at acknowledging their achievements, inspiring excellence, and aligning their career trajectory with organizational objectives.

1. **Performance evaluation:** To identify candidates for promotion, conduct routine performance evaluations to gauge employees' skills, achievements, and capacity for advancement. Employ performance metrics, feedback from supervisors and peers, and goal achievement data to recognize top-performing employees.
2. **Set Clear criteria:** Develop precise criteria and guidelines for promotions, focusing on performance, skills, experience, leadership capabilities, and alignment with organizational values and objectives. Foster transparency in the promotion process to uphold fairness and equitable opportunities for all employees.
3. **Communicate expectations:** Ensure clear communication of promotion eligibility criteria, expectations, and timelines to employees. Offer insights into areas for improvement and avenues for development to support employees in readying themselves for advancement.
4. **Career development plans:** Partner with employees to formulate tailored career growth roadmaps delineating goals, training needs, and skill enhancement efforts for promotion.

Recognition and Rewards

Show appreciation and incentivize employees for their valuable contributions and successes by granting promotions, salary increments, bonuses or supplementary benefits. Publicly commend their dedication to inspire team spirit and drive.

1. **Promote internal mobility:** Promote internal mobility by giving preference to internal candidates for promotions.
2. **Feedback and support:** Offer consistent feedback, coaching, and assistance to employees to assist them in tackling challenges, improving performance, and honing the skills essential for promotion. Foster transparent communication and resolve any issues or obstacles hindering their progress.
3. **Merit-based promotions:** Base promotions on merit and performance rather than seniority or tenure alone. Evaluate employees objectively and consider their contributions, skills, potential, and alignment with organizational goals when making promotion decisions.
4. **Document promotion process:** Maintain accurate records of the promotion process, including performance evaluations, development plans, feedback, and promotion decisions. Documenting the process ensures transparency, consistency, and accountability.
5. **Follow-up and evaluation:** Follow up with promoted employees to ensure a smooth transition into their new roles. Evaluate the impact of promotions on employee engagement, retention, and overall organizational success to refine future promotion strategies.

By following these steps, organizations can promote employees effectively, foster a culture of growth and development, and align talent management strategies with business objectives.

Satisfaction Rate

Meaning

When we talk about 'satisfaction rate', we are referring to a gauge used to quantify how satisfied individuals or groups are with a particular product, service, experience or scenario. This metric reflects the proportion of satisfied respondents compared to the total number of respondents included in the survey or evaluation.

The satisfaction rate is typically expressed as a percentage and can be calculated using the following formula:

$$\text{Satisfaction Rate} = \frac{\textbf{Total Number of Respondents}}{\textbf{Number of Satisfied Respondents}} \, x \, 100\%$$

Indicators of Satisfaction Rate Means in Practical Terms

Sl.No.	*Level of Satisfaction*	*Indicators*
1.	Customer satisfaction	High satisfaction indicates happiness of the customers leading to loyalty, positive and repeat of business.
2.	Employee satisfaction	Indicates positive work environment, job satisfaction and motivation.
3.	Service quality	Indicates services meet the expectation.
4.	Feed back	Indicates strength, weakness to take appropriate action.
5.	Bench marking	Indicates achievement, task, market trend and areas of competition.
6.	Business growth	Indicates feeling of customer, loyalty, attracting new customer and brand reputation.

Overall, the satisfaction rate is a valuable metric for assessing stakeholder satisfaction, identifying opportunities for enhancement, and driving continuous improvement across various aspects of business and service delivery.

Features of Promotion

a. Promotion is a reward for consistently good performance of the employee.
b. It enhances employee morale and job satisfaction.
c. It involves higher responsibilities, better job status and increase in salary.
d. Provides opportunities for career advancement.
e. It helps the organisation to attract and retain good performers.

The Main Purposes of Promotion

a. To recognize and reward the efficiency of an employee.
b. To attract and retain the services of qualified and competent people.
c. To increase the effectiveness of the employee and of the organization.
d. To motivate employees to higher productivity.
e. To fill up higher vacancies from within the organisation.
f. To impress upon those concerned that opportunities are available to them also in the organization if they perform well.
g. To build, loyalty, morale and sense of belongings in the employees.

Limitations of Promotion

Though promotion is important for the employee and the organisation, it creates certain problems which are as follows:

1. **Some employees refuse promotion:** Employees generally tend to accept promotions. There are several incidents where employees refuse promotions because transfer to an unwanted place, incompetent to carry out the job, delegation of unwanted responsibilities and family problem.
2. **Promotion disappoints some employees:** Some employees who are not promoted will be disappointed when their colleagues with similar qualifications and experience are promoted.

G. Job Rotation

Definition

Job rotation is the systematic movement of employees from one job to another within the organization to achieve various human resources objectives such as orienting new employees, training employees, enhancing career development, and preventing job boredom or burnout.

Job Rotation means laterally moving an employee to different positions, departments, or geographical locations to professionally develop the employees by exposing them to new knowledge, skills, and perspectives.

Four Types of Job Rotation

1. **Horizontal job rotation:** Employees of the same hierarchical level or department are moved across different roles or positions to help broaden their skills and knowledge of a specific functional area
2. **Vertical job rotation:** Employees are moved to either higher or lower hierarchical levels to gain a broader understanding of the responsibilities of various levels of the organization.
3. **Lateral job rotation:** Employees are moved across different departments or functional areas for cross-functional experience and develop a holistic view of the organization's operations
4. **Geographical job rotation:** Employees are relocated to different locations or regions to be exposed to diverse markets, locations, or regions and enhance their global perspective.

Five Methods of Job Rotation

1. **Job enlargement:** It provides a broad spectrum of employee roles and responsibilities and increases their sense of challenge and fulfilment by adding to the current role and expanding its scope.
2. **Job enrichment:** The employee's current job role is enhanced with additional autonomy and responsibility. It helps to provide a higher level of job satisfaction by making the job more meaningful and fulfilling.
3. **Job simplification:** It aims to simplify complex tasks to reduce the burden on employees and streamline the job responsibilities. This ultimately helps to enhance efficiency and productivity.

4. **Cross-training:** Employees are trained in multiple roles or functions to acquire a broader skill set. This helps improve flexibility and enables employees to cover for each other during absence, so the fluctuations don't affect the quality of work.
5. **Temporary assignments:** It involves temporarily assigning employees to different roles to help them gain exposure to new areas of the organization and develop a broader understanding of the organization.

Job Rotation - Training Methods

The following are the 5 training methods for job rotation:

Training Methods	*Description*
1. Shadowing	Employees observe and learn from those already holding the role that they will rotate into.
2. Mentoring	Employees undergoing job rotation and experienced employees are paired together, facilitating guidance, support, and knowledge transfer.
3. On-the-job training (OJT)	Employees learn by performing tasks in the actual work environment. In addition, employees receive hands-on training and guidance from trainers during their transition into newer roles.
4. Cross-functional training	Employees receive formal training on skills and knowledge required for the newly rotated roles which they receive it from internal trainers, external experts, or learning modules.
5. Job rotation workshops	Organizations promote the idea of job rotation by organizing workshops cantered around communication, problem-solving, decision-making, and tools to streamline the process.

To make the training methods more effective, following 6 key steps may be adopted;

1. Identify training objectives
2. Assess employee skills and interests
3. Design a rotation plan
4. Collaborate with managers and supervisors
5. Communicate with employees
6. Evaluate and monitor progress

Job Rotation in HRM - Advantages and Disadvantages

Advantages	*Disadvantages*
Reduces burnout to overcome monotony and boredom.	Personal preference and career aspirations may create resistance
Reduces absenteeism	Insufficient training and education for the new roles may create problem
Improve performance	Challenges of physical relocation
Increased innovation	Resistance to sharing machines or equipment
Multitasking and flexibility	Reluctance to learn and adapt
Improved interpersonal relationships	

How to make the Job Rotation Successful?

To make the job rotation successful, the following factors should be considered.

1. Firm characteristics	i. Clear objectives ii. Supportive culture iii. Effective communication
2. Workforce characteristics	i. Competency ii. Employee engagement. iii. Continuous feedback
3. HRM practices	i. Training and development ii. Performance management iii. Succession planning
4. Supportive infrastructure	i. Resources and support ii. Seamless transition iii. Performance evaluation

H. Reward

Meaning

Rewards and recognition are important tools that can be used to motivate employees and improve their performance. Rewards can include things such as bonuses, raises, or special privileges, while recognition can come in the form of words of appreciation or awards.

Reward is an incentive plan to reinforce the desirable behaviour of workers or employers and in return for their service to the organization. Rewards can be monetary in the form of salary or non-monetary in the form of awards for some special services to the company or simply giving an employee a work which he enjoys doing. The primary objective of organizations in giving rewards is to attract, maintain and retain efficient, high performing and motivated employees.

Importance of Rewards

There can be various types of rewards an organization can give its employees like money, grade, performance based incentive or performance pay, increment, gift cards, recognition or awards, profit sharing, holiday packages, medical coverage, promotion, bonuses *etc.*

Types of Rewards in HRM

Rewards can be of two types:

1. **Intrinsic:** They are incentives which satisfy an employee internally. Only money is not enough to motivate people and it is important to make people realize their contribution to the organization matters. That motivates employees internally
2. **Extrinsic:** They are mostly tangible incentives like pay, advancement, recognition, time off *etc.* Extrinsic rewards the ones which are more sought after by employees as it can improve the motivation and job satisfaction in the short term and addresses immediate issues.

There are a number of different ways to use rewards and recognition, and the best approach will vary depending on the individual company and its employees. However, some general tips include:

1. Establish clear goals and objectives for employees, and ensure that rewards are tied to performance.
2. Make sure that rewards are meaningful and relevant to employees.
3. Rotate rewards and recognition so that employees don't become bored or complacent.
4. Be creative with recognition, and use a variety of methods to show appreciation.
5. Keep track of employee satisfaction and performance, and make changes to the rewards and recognition program as needed.

Benefits of Rewards and Recognition

There are many benefits of rewards and recognition programs in organizations. Recognition motivates employees to do their best and feel appreciated, while rewards encourage employees to meet or exceed specific goals. Furthermore, rewards and recognition help to build teams and create a positive work environment. Recognition also strengthens relationships between employees and managers, while rewards help to develop a sense of camaraderie. Finally, rewards and recognition can improve employee satisfaction and reduce staff turnover.

Difference between Rewards and Recognition

Reward	*Recognition*
Rewards are the gifts and awards given to employees for their accomplishments	Recognition is the appreciation or praise for their accomplishments
Rewards have some monetary value attached to them	It is intangible
They are economical appreciation	Recognition is emotional

How to Build a Successful Employee Rewards and Recognition Program?

A successful reward recognition program should have the following:

i. **Personalized rewards**: Every individual should be able to get a tailor-fit incentive program according to their desires.

ii. **Clear messages**: The points like the program's purpose, team building, training, *etc.*, should be clear.

iii. **Should link rewards with business outcomes**: The employee should understand the connection between the overall goals of the company and their work.

11.4 Development System - Induction, Training, Job Enrichment, Self-learning Mechanisms, Potential Appraisal, Succession Development, Counselling, Mentor System, Self-renewal System, Survey and Action Research

A. Induction

Definition: Induction is the process for welcoming newly recruited employees and supporting them to adjust to their new roles and working environments. Starting a new job can be a stressful experience and new employees need help to settle in.

In the context of Human Resource Management (HRM), "induction" refers to the process of introducing and orienting new employees to their organization, their job roles, and the company's culture and values. The primary goal of company induction is to help new employees integrate into the organization smoothly, understand their job responsibilities, and become productive members of the team. HR induction meaning and practices differ for different organizations.

Objectives of Induction

1. **Effective role clarification:** To ensure that new employees understand their job responsibilities and expectations from the outset, reducing ambiguity and improving job performance.
2. **Familiarization with organizational culture:** To acquaint new hires with the company's values, beliefs, and cultural norms, promoting alignment with the organization's ethos and fostering a positive work environment.
3. **Safety and compliance awareness:** To educate employees about safety protocols, legal compliance requirements, and company policies, ensuring a safe and compliant work environment.
4. **Team integration and relationship building:** To facilitate connections with colleagues and supervisors, fostering a sense of belonging, teamwork, and cooperation among new and existing employees.
5. **Efficient productivity:** To expedite the transition to full productivity by equipping employees with the necessary knowledge and skills, reducing the time it takes for them to become valuable contributors to the organization.

The other objectives are, cultural integration, awareness of legal aspects and Enhanced Brand Image.

Purpose of Induction

The purpose of employee induction is to ensure that new hires have the knowledge, tools, and support needed to become effective and engaged members of the organization, benefiting both the employee and the company. Induction helps new employees integrate into the organization and their new roles seamlessly, reducing the stress and uncertainty often associated with starting a new job.

How does Induction Work?

Employee induction is a structured on boarding process that helps new hires integrate into an organization. It typically begins before the employee's start date, with administrative tasks like paperwork and background checks.

On the first day, new employees receive a warm welcome and are introduced to the company's culture, policies, and their job roles. They undergo training, meet their colleagues, and become familiar with the workplace. Throughout the induction period, employees receive ongoing support and guidance, including feedback sessions. The induction process ensures that new employees quickly become productive, aligned with the company's values, and feel part of the company.

Types of Induction

1. **Formal induction:** This is a structured and organized on boarding process where newly recruited employees follow a pre-defined schedule and curriculum. It typically includes presentations, training sessions and orientation activities.
2. **Informal induction:** Informal induction relies on a more casual, ad-hoc approach. New employees may learn about their roles and the organization through interactions with colleagues and supervisors without a structured program.
3. **Individualized induction:** This type of induction is tailored to each newly recruited employee's unique needs and requirements. It considers their prior experience, skills, and specific job role, offering personalized training and support.
4. **Group induction:** In a group induction, several new employees go through the on boarding process together. It can be a cost-effective way to provide information and build camaraderie among new hires.
5. **Virtual/remote induction:** With the rise of remote work, virtual induction takes place online. It involves video conferencing, webinars, e-learning modules, and digital resources to on-board new employees, even when they are not physically present in the workplace.
6. **Role-specific induction:** This job induction type focuses on providing job-specific training and information. It equips new employees with the understanding of her/his job, skills, and knowledge necessary for their particular positions.
7. **Cultural induction:** Cultural induction emphasizes introducing new employees to the organization's culture, values, and norms. It helps them align with the company's ethos and work effectively within its cultural framework.

What is the difference between Employee Induction and Employee Orientation?

Employee induction and employee orientation are related but distinct processes in the on boarding of new employees. Employee induction refers to the entire on boarding process, encompassing a more comprehensive approach that aims to integrate new employees into the organization.

On the other hand, employee orientation is a subset of the induction process, focusing primarily on introducing new employees to the organization's culture, values, workplace facilities, and essential administrative procedures.

What is the Process of Employee Induction?

The process of induction in HRM can be divided into three phases: Day one induction, week one induction, and month one induction. The checklist for all these phases is mentioned below.

Day one induction checklist	*Week one induction checklist*	*Month one induction checklist*
On the first day for employees, ensuring the feel enthusiastic and engaged is important. Day one checklist are team introductions, buddy introductions and workplace tour followed by welcome lunch.	During week one, new joiners to spend time with their team, understand the task requirements, and what are expected from them. Activities are, organization overview, Introduction, payroll introduction, workplace health and safety induction and lastly the performance overview.	In first month after joining, employees get clarity on their job roles and how their department or team respond to different job responsibilities.

Benefits of an Induction Programme

1. It'll save on time and resources.
2. It's an opportunity to get creative incentives. Instead of just treating an induction as a tick-box exercise.

B. Training

Definition

Training and development in Human Resource Management (HRM) refers to a system of educating employees within a company. It includes various tools, instructions, and activities designed to improve employee performance. It's an opportunity for employees to increase their knowledge and upgrade their skills.

Training and development play an important role in positioning a company's success. The wake of the advancement of technology has resulted in tight competition, and a rise in customers' expectation of quality and service at lower costs. It is also becoming more important globally to prepare workers for new jobs.

Training and Development

The need for training and development in India has risen over time due to the changing workplace dynamics and the accelerated advancement in technology. Businesses have to constantly evolve to stay ahead in the competition.

Employee Training and Development

Employee Training and development is an ongoing process where companies help or train their employees acquire the knowledge and skills needed to perform better at their job roles as well as reach their full potential and give their best performance. Training employees is teaching them how to perform a specific task or procedure.

Six Training and Development Processes

The right training and development programs can improve employee performance, increase productivity, and help organizations remain on top of the table. The HRM team in an organization plays a vital role in developing programs that benefit employees.

The main process involved in training and development are as follows-

Process	*Activity*	*Method*
1. Need assessment	Identification of knowledge, skill and competency needed for the job	Gap analysis collecting data through survey, interview and Focus Group Discussion. Gap between existing and requirement.
2. Design and development	Designing and developing learning goal	Training methods, Teaching learning materials, formulation of training strategy and modules.
3. Delivery	Implementation of training program	On-line, off line, on the job training, training environmental facility, equipment.
4. Evaluation	Impact of training program	Evaluation if methods, tests, quiz, observation, survey
5. Implementation	Rolling out training program	Communication of purpose, observation, benefits of training followed by documentation
6. Post training support	Reinforce learning	Job application observation

Importance of the Training and Development

1. Skill enhancement
2. Increased productivity and efficiency
3. Adaptation to technological advancements
4. Employee engagement and retention
5. Succession planning and leadership development

6. Improved employee morale and motivation
7. Improved customer satisfaction

Training and Development Methods

There are 6 training and development methods.

1. **On-the-job training:** Trainees are placed on a regular job and are taught the necessary skills on the go under the guidance of a supervisor or instructor. The three techniques for the on-the-job training are-
   - ✰ Coaching- A person to person interaction.
   - ✰ Mentoring- junior employees learn under the guidance of seniors.
2. **Committee assignments:** The trainee acts as a part of a committee, observes activities and investigates organizational problems.
3. **Vestibule training:** In a vestibule setup, with the actual job simulations, expert trainers are employed to use necessary equipment and machines used in the workplace.
4. **Apprenticeship training:** Theoretical and practical training is given to trainees under government established institutes and are also paid a stipend during the periods.
5. **Classroom training:** Training is provided in company classrooms and educational institutions. Group discussions, case studies and audio video visual aids are all part of the training mode.
6. **Internship training:** It is a joint effort of an educational institution and business firms to provide training to candidates making sure there is a right mix of theory and practical knowledge making the job ready.

Objectives of Training and Development

The purpose of training and development is:

1. Employee well-being, promotion of growth and career advancement.
2. Can lead to the production of future leaders.
3. Results in low attrition rates.
4. Employee satisfaction leads to better employee performance and greater organizational productivity.

Benefits of Training and Development

The impact of training and development are as follows:

1. Higher employee satisfaction and morale.
2. Reduced employee turnover.
3. Hard working employees.
4. Improved efficiency, resulting in financial benefit.

5. Increased innovation and creativity.
6. Improved engagement.
7. Better problem-solving skills.
8. Improved retention and loyalty.

Types of Training and Development

Essentially Employee training and development programs can be classified as:

1. Orientation training
2. Leadership training
3. Technical training
4. Compliance training
5. On boarding training
6. Product training
7. Soft skills training
8. Sales training

Difference between Training and orientation

Basis	*Training*	*Development*
1. Meaning	A process in which employees are given an opportunity to develop skills and knowledge that is job relevant	An educational process that concerns the growth of the individual overall
2. Time limit	Short term	Long term
3. Orientation	Job orientation	Career orientation
4. Suitability	For Technically inclined staff	For Managerial staff
5. Trainee level	Non-managerial	Managerial
6. Skills	Technical skills only	Technical, soft, and conceptual skills
7. Objective	Improve job performances	Prepare employees for future career challenges

C. Job Enrichment

Job enrichment is a strategy used to motivate employees by giving them increased responsibility and variety in their jobs. The idea is to allow employees to have more control over their work. By doing so, one can tap into their natural desire to do a good job and contribute to the overall goals of the company.

Definition

Job Enrichment is the process of enhancement of a job by increasing the level of responsibility and autonomy of employees. It provides employees with challenges and more accountability to instil a feeling of achievement and growth.

Organizations use job enrichment in HRM (Human Resources Management) as a strategy to make jobs more meaningful and engaging for employees. The job enrichment process expands job responsibilities, grants employees autonomy, and provides growth and development opportunities. This aims at higher employee satisfaction, motivation and increases the overall performance of the organization.

According to author, David Bennett, job enrichment is an approach to structuring jobs that provides with more self-actualization than receive during job enlargement or job rotation. His theory involves exploring jobs vertically more than horizontally

1. The purpose of job enrichment is to enhance job content to satisfy employees' psychological and social needs, ultimately increasing job satisfaction.
2. It involves providing opportunities for employee motivation, such as suggestion programs, job rotation, information sharing, quality circles, and classroom training.
3. The results of the study suggest that job enrichment positively impacts job satisfaction and does not lead to increased work intensity or decreased job security.

Taylorist Jobs Vs enriched Jobs

According to Fredrick Taylor (1947), job design is a scientific problem of finding the most efficient way to divide work into specific tasks. In 'Taylorist' jobs, workers have highly specialized roles and are discouraged from experimenting, being innovative, or deviating from the prescribed way of completing tasks. The focus is on maximizing efficiency rather than encouraging creativity or flexibility.

In today's world, 'Taylorism' has gained a bad reputation due to its stringent features. Taylor's job design principles made way for the automation and standardization of processes. Although this is universal in the 21st-century workplace culture, the upcoming generations crave variety more than ever.

Difference between Taylorist Jobs and Enriched Jobs highlighting the key characteristics of each.

Taylorist Jobs	*Enriched Jobs*
i. Highly specialized	i. Increased responsibility
ii. Limited task variation	ii. Greater autonomy
iii. Efficiency-focused	iii. Encourages creativity

Taylorist Jobs	*Enriched Jobs*
iv. Minimal input by individuals	iv. Opportunities for growth
v. Prescribed work methods	v. Flexible with methods, challenging and fulfilling

How to Enrich Jobs - Practical Ways

As most leading HRM (Human Resources Management) strategies come up as solutions to problems in the workplace, job enrichment was conceptualized for some problems too.

Some of these problems include,

i. Uninterested employee performance
ii. High rate of absenteeism
iii. Lack of engagement and satisfaction
iv. High rate of employee turnover

Enriching jobs is taken up as a management initiative as a solution to the above problems. There is no single defined method of enriching jobs. The method or technique used is dependent on the situation or the circumstance. According to Tripathi in 'Employee Motivation and Job Enrichment', the 10 most common methods followed while enriching jobs are,

1. Task rotation and expanding skills and responsibilities.
2. Allowing employees to have more control and flexibility in how they perform their work.
3. Giving employees complete tasks instead of highly specialized ones.
4. Reducing excessive supervision while holding individuals or groups accountable for their performance.
5. Involving employees in setting targets, standards, and decision-making processes.
6. Providing employees with the necessary information to monitor their own performance.
7. Encouraging employee participation in planning, innovation, and reviewing outcomes.
8. Introducing new and challenging tasks for employees.
9. Assigning individuals or groups specific projects to increase their expertise and responsibility.
10. Allowing employees to make decisions during problem situations rather than relying solely on their superiors.

Job Enrichment Models

We shall study how these models provide strategies for redesigning jobs by incorporating elements such as increased autonomy, skill variety, task significance, and feedback. By applying these models, companies aim to create enriched jobs to foster a sense of accomplishment, growth, and engagement among employees.

Job Enrichment through Job Crafting Model

Job Crafting is a process in which employees redesign their jobs by altering their job responsibilities and duties in a way that makes them more engaged and satisfied. It is seldom a one-time event. It is a continuous process that employees go through during their tenure in a job role. The process takes place in 3 main stages.

1. Motivation: Finding the motivational factors to craft and redraft the job.
2. Identification: Identifying and discovering the opportunities available to redesign the job.
3. Outcomes: Observing outcomes after crafting new models for the job.

Job Enrichment through Job Characteristic Model

The Job Characteristics Model, developed by Hackman and Oldham in the 1970s, offers recommendations on how to enhance jobs within organizations. The model suggests five key characteristics for effective job design.

1. **Skill variety:** Using a number of different skills and talents to complete tasks in a job.
2. **Task identity:** Completing a task as a whole, from beginning to end with a definite outcome.
3. **Task significance:** The impact of the job on the team, organization, or the world at large.
4. **Autonomy:** Having the freedom and independence to decide on the schedule and procedure to complete tasks on the job.
5. **Job feedback:** Obtaining information and feedback from peers or managers about the effectiveness of carrying out the job.

According to the model, these characteristics impact motivation, job satisfaction, and performance. The model also includes additional factors such as the meaningfulness of the work, sense of responsibility, and knowledge of results as important influences on job outcomes.

The Job Characteristics Model produces a single index called Motivating Potential Score (MPS) that measures the motivating potential of a job. It can be obtained from the above 5 characteristics using this formula:

$$\textbf{Motivational Potential Score (MPS)} = \frac{\textbf{SV + TI + TS}}{\textbf{3}} \times \textbf{Autonomy} \times \textbf{Feedback}$$

(SV: Skill variety, TI: Task identity, TS: Task significance)

Advantages and Disadvantages of Job Enrichment

Advantages	*Disadvantages*
Interesting and challenging job responsibilities	Insufficient knowledge to take decisions produces the wrong attitude in the workplace
Better decision making	Some employees experience overwork
Increased job satisfaction	Potential resistance to change
Potential for personal and professional growth	Need for additional support and training
Identification of potential leaders and managers	More power to employees can lead to ego problems
Points out employee needs and requirements in a different level	Internal dissatisfaction cannot be addressed with enrichment
Reduced workload for superiors and managers	Unsuitable for job roles with already enough independence and responsibilities

D. Self-Learning Mechanisms

One of the most basic learning mechanisms is memorization. This approach captures a sequence of steps or a response to a specific set of conditions and then, when the same task or set of conditions is again encountered, the memorized responses are applied.

Definition

According to Malcom Knowles, it is a mechanism through which people take the initiative to diagnose their learning needs, devise learning strategies, define human and material resources for learning, and determine learning outcomes, with or without the assistance of others.

Advantages of Self-Learning

1. **Development of strong problem-solving skills**: Self-learning allows to recognise issues and easily check on own for successful solutions. This may be through friends, the internet or by discovery. Whatever the case, obstacles and hurdles do not destroy desire to get things done, but instead, give new opportunities to know how to systematically search for solutions.
2. **It is a stress-free learning process**: The self-learning process puts a great focus on the process rather than on the learning outcome. Also, within a certain time, there is no obligation to learn the necessary material, with the ultimate aim being the achievement of high grades. When we want to

learn, choose what we want to learn and how to learn. The net result is material internalisation, which is why self-learners appear to internalise what they learn more than those who engage in guided learning.

3. **Gaining of additional skills in the process**: These are skills that can be applied everywhere. In particular, individuals who are good at self-learning have opportunity to; learn many other skills in the process of self-learning.
4. **The learning experience becomes more meaningful**: Self-driven learning derives from a personal desire to develop new skills. The desire to find more knowledge about a subject and use the information for a specific reason is motivated by it. It becomes more meaningful to the learners.
5. **Learning is now led by curiosity**: A big part of learning is to satisfy some degree of interest, and when there is interest, learning happens naturally satisfying curiosity to know and learn.
6. **Own choice to select mode of learning**: When we plan to learn, from YouTube videos, webinars, books or any other form of learning is at our disposal. We pick one that suits best and one that we find most engaging with all these modes of learning.
7. **It creates a sense of responsibility**: Self- learning compels the person to use them to make their notes and study. Within a person, an idea of obligation is constructed. They begin to feel responsible for what they do, which in turn also leads to character building.
8. **Easy language**: Self- learning requires attempts to read, understand, make notes and then learn them. The learners select easy language to learn.
9. **Builds confidence**: The best way to raise confidence is self- learning. Self- learning satisfies the independence of an individual.
10. **Score well**: Being self-taught allows an individual to score better. One where an individual is competing with himself is a healthy rivalry.

Disadvantages of Self-Learning

1. No self-discipline
2. No face-to-face interaction
3. Lack of input from trainers
4. Slow evolution
5. Good e-learning is difficult to do
6. Lack of transformational power
7. No peripheral benefits

The Importance of Self-Learning

1. It inculcates discipline and boosts self-esteem
2. It improves problem-solving
3. It makes learning relevant
4. It is a customisable experience
5. It helps one develop problem-solving skills
6. It grows self-knowledge
7. Gives room to an individual to have multiple skills

How to Start Self-Learning?

1. Step out of comfort zone
2. Establishing the goals
3. Nurturing strengths
4. Think about career goal
5. Be honest with self
6. Get guidance from someone trust
7. Evaluate outcomes

E. Potential Appraisal

Potential appraisal is a future: Oriented appraisal whose main objective is to identify and evaluate the potential of the employees to assume higher positions and responsibilities in the organizational hierarchy. Many organisations consider and use potential appraisal as a part of the performance appraisal processes

It can be measured through:

1. Self-appraisals
2. Peer appraisals
3. Superior appraisals
4. MBO (Management by Objectives)
5. Psychological and psychometric tests
6. Management games like role playing
7. Leadership exercises

F. Succession Development

Definition

Succession development is a method used to identify and develop emerging leaders so that they can take control in time of need and keep a company afloat and unaffected. A succession planning development plan is the stepping-stone to creating a leadership pipeline so that a business continues to move forward, even amid unanticipated changes.

Succession development ensures an improved understanding of the talent pipeline available internally while creating awareness of the current individuals' strengths.

Succession Planning

It involves identifying and training future leaders from internal sources who will take forward business making proper use of resources and assets. Succession plan enhances an organization's capabilities by:

1. Pinpointing vital roles and potential gaps
2. Identifying essential competencies for continuity
3. Documenting critical information to facilitate knowledge transfer
4. Prioritising employee development to meet business demands

How to Develop a Succession Plan?

Every organization has its own people requirements and business goals. This means that each organization will have its own way of developing a succession plan. Research indicates that there are 8 steps to develop effective succession plan.

Step 1 - Define objectives of succession plan

Succession plans need to have a clear objective or goal. This ensures that the organization's resources are not wasted in formulating and implementing the succession plan.

Some objectives of a succession plan can be:

1. A succession plan can be made specifically for finding successors for C-suite positions such as CEO, CFO, and Heads of different departments.
2. A succession plan can be created to find the right future fit for middle-level leadership and managerial positions.
3. A succession plan can be made for future talent needs based on the company's growth and expansion plans or to accommodate changing market scenarios.
4. A succession plan can be created especially to find replacements for retiring board members or to fill in new positions.

Step 2 - Assess current leadership situation

After objectives are set it is necessary to assess internal potentials. During the evaluation process:

1. Determine suitable persons to step into the position.

2. Consider the strengths and weaknesses of persons in terms of abilities to inspire,
3. Assess whether they are equipped to assume their roles or need further training
4. Measure how current leaders' objectives align with the broader business goals,
5. Identify the key stakeholders who interact with your leadership team and assess how well leaders manage these relationships.

Step 3 - Identify the potential successors

Identifying potential successors involves a systematic approach to talent assessment and evaluation. This step encompasses a thorough analysis of skills, competencies, and leadership potential across various organizational levels. Consider these strategies to pinpoint individuals ready to assume critical roles:

1. **Talent mapping** - Developing a comprehensive inventory of high-potential employees involves considering their current roles, skill sets, and historical performance.
2. **Performance review** - Reviewing historical performance evaluations help to identify consistent high performers who exhibit promising leadership potential.
3. **Succession readiness** - Assessing of the readiness of potential successors through factors such as adaptability, learning agility, communication skills, and problem-solving abilities. Review may be done on the following aspects
   a. Talent mapping
   b. Performance review
   c. Succession readiness

Step 4 - Formulate a development plan for successors

Make a list of potential successor. The key elements are,

1. Skill enhancement
2. Mentorship opportunities
3. Cross-functional exposure

Step 5 - Communicate succession plan to stakeholders

Transparently sharing the plan's details and benefits fosters support and engagement among stakeholders. The strategically communicate plan is to focus on,

1. Transparency
2. Open dialogue.
3. Managing expectations Step

Step 6 - Align with top management

The active engagement of top management is pivotal in both formulating and executing a successful succession plan. The following points are helpful at this stage.

1. Strategic alignment
2. Resource allocation
3. Championing leadership development

Step 7 - Regularly review and update succession plan

To maintain the effectiveness of succession plan, engage in periodic reviews and revisions. Evaluate the viability of planned changes and their readiness for implementation when required. The following points are helpful at this stage.

1. Scheduled assessments
2. Emerging.
3. Skill evolution

Step 8 - Execute succession plan

To ensure a successful implementation, consider these strategic approaches:

1. Cultural integration
2. Training integration

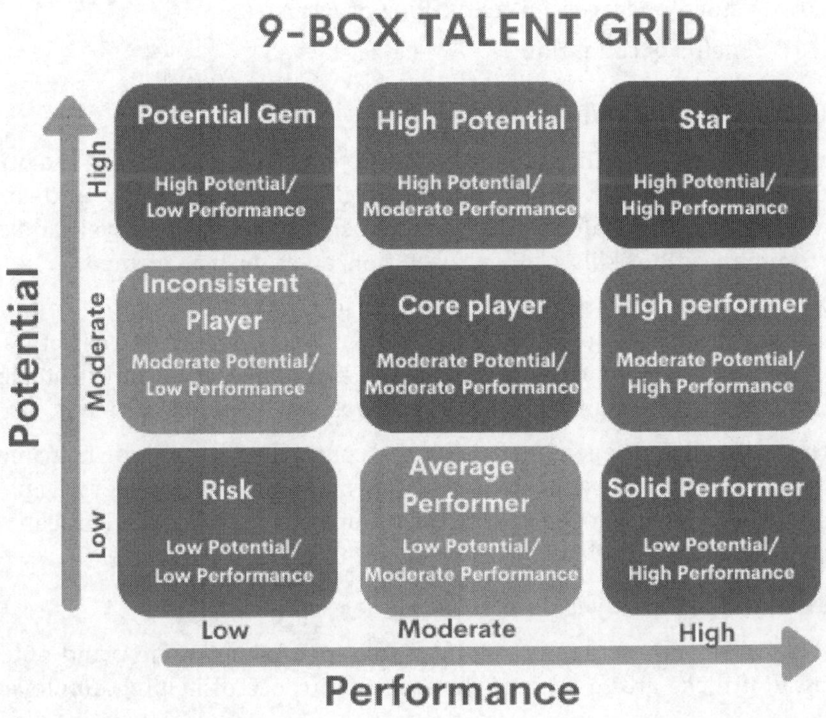

The 9-Box Grid is a popular tool in succession planning. It is often used to assess leaders based on performance and potential. It categorizes leaders into different quadrants, such as high performers with high potential (potential successors), solid performers, and those needing development. Utilize this grid to represent leadership assessments visually and to identify potential successors.

G. Counseling

Counselling is described as the help provided by the supervisor to the subordinates in analysing their performance and other behaviours on the job, in order to improve, their performance. Counselling is also used sometimes in the sense of coaching and reviewing one's performance

In this section we shall be dealing with counselling in the following aspects.

1. Introduction to Employee Counselling
2. Meaning of Employee Counselling
3 Concept
4. Need and Objectives
5. Types of counselling
6. Process of counselling
7. Performance achievement8. Functions
9. Qualities of a counsellor
10. Factors for successful counselling program
11. Benefits of counselling

1. Introduction to Employee Counselling

Behavioural counselling in human resource management (HRM) is a process that involves working with employees to identify and modify undesirable behaviours that may be affecting their work performance. It effectively addresses poor communication skills, conflict resolution, and time management.

Behavioural counselling can take various forms, including individual counselling sessions, group counselling sessions, or a combination of both. These counselling sessions aim to help employees develop new skills and strategies for improving their behaviour and encourage a positive work environment.

One of the critical benefits of behavioural counselling is that it can help improve workplace productivity and morale. By addressing negative behaviours, employees are more likely to work together effectively and be more engaged. This can result in increased job satisfaction and reduced absenteeism and turnover.

2. Meaning of Employee Counselling

Employee counseling is a vital part of performance review and potential appraisal, if these are to achieve their basic purpose of helping employees to improve and develop. Unless carefully and sensitively handled, employees may become more dissatisfied after the counseling than before.

Definition: Employee counseling is a method of understanding and helping individuals who have technical, personal and emotional adjustment problems interfering with their work performance.

The personal adjustment of an employee is often improved when he is able to obtain information that he needs for reaching an intelligent solution to his problems or when he is able to express his feelings without fear of reprisal. Counseling is a sophisticated form of communication in which attempts are focused to maintain constructive attitudes.

3. Concept

Counseling is a two-way process in which a counselor provides help to the workers by way of advice and guidance. There are many occasions in work situations when a worker feels the need for guidance and counseling. The term 'counseling' refers to the help given by a superior to his subordinate in improving the latter's performance. There are three aspects in explaining concept of counselling.

i. Counselling deals with emotional problems.
ii. Counselling involves discussion *i.e.*, it is an act of communication.
iii. The general objective of counselling is to understand and/or decrease an employee's emotional disorder

4. Need and Objectives

The need for employee counseling is,

i. To come out of problem and find new way to work
ii. To know as to how much the employer care for the employee.
iii. Need to identify the work related problems and the poor performance.
iv. Need to increase the productivity of employee

Objectives of counseling are as follows:

i. Counselling is an exchange of ideas and feelings.
ii. It is concerned with both personal and work problems.
iii. Counselling is performed by both professionals and non-professionals.
iv. Counselling is confidential.
v. It tries to improve organizational performance.
vi. Encouraging employees to set goals for further improvement

The concept of mentoring relates to emotional support and guidance usually given by a senior person to a younger one known as protégé. Both counseling and mentoring involve help and support by a senior person to a younger one.

The three basic ingredients of the process are:

i. Communication,
ii. Empowering and
iii. Helping.

5. Types of Counselling

Effectiveness of counseling largely depends on the methods and techniques as well as the skills used by the counselor. Methods and techniques of counseling change from person to person and from situation to situation.

Normally employee counselling involves the following types:

i. Directive counselling
ii. Non-directive counselling
iii. Cooperative counselling
iv. Participative counselling
v. Desensitization
vi. Catharsis
vii. Insight
viii. Developing the new patterns

Type 1 - Directive counselling

It is full counselling. It is the process of listening to an employee's problem, deciding with the employee what should be done and telling and motivating the employee to do it. It centers on the counselor. The counselor, after hearing the problems of an employee, decides what should be done and gives advice and suggestion to him to resolve the problem. But directive counseling seldom succeeds, as people do not wish to take up advice normally, no matter how good it might be.

Type 2 - Non-Directive Counselling

It is the process of skillfully listening to the emotional problems of an employee, understand him/her and determine the course of action to be adopted to resolve his problem.

Type 3 - Cooperative Counselling

It is the process in which both the councilor and client mutually cooperate to solve the problems of the client. It is not neither wholly client centered nor wholly counselor centered but it is centered both counsellor and client equally. Among the three from of counseling, the advice offered in directive counseling considers the surface crises; the nondirective counseling goes to the underlining cause, the real crisis that leads the employee to understand his problem. It is thus suggested that nondirective to counseling is, probably, the best among the three forms.

Type 4 - Participative Counselling

Both directive and non-directive methods suffer from limitations. While the former is often not accepted by independent employees, the latter needs professionals to operate and hence is costly. Hence, the counseling used in most situations is in between these two. This middle path is known as participative counseling.

Type 5 – Desensitization

According to Desensitization, once an individual is shocked in a particular situation, he/she gives himself/herself no chance for the situation to recur. This method can be used to overcome avoidance reactions, so as to improve the emotional weak spots.

Type 6 - Catharsis

Discharge of emotional tensions can be called catharsis. A Catharsis is an emotional discharge through which one can achieve a state of moral or spiritual renewal or achieve a state of liberation from anxiety and stress.

Type 7 – Insight

Founded by Sigmund Freud, psychoanalysis or insight delves deep into an employee's past and brings to light past experiences and current unconscious thoughts and behaviours of the employee, that are believed to be the cause of their current problems.

Type 8 - Developing the new patterns

Developing new patterns becomes very often necessary when other methods to deal with weak spots remain ineffective. In order to develop new, more satisfying emotional reactions, the individual needs to expose himself to situations where he can experience positive feelings.

6. Process of Counselling

Counseling, basically, is to help the employee grow and develop in the organization. An effective counselor is one who helps employees to make them aware of their strengths and weaknesses and to improve upon them. Through mutuality, he/she helps the employee to develop providing them the proper emotional climate.

The counseling process has three important phases:

i. **Rapport building:** Rapport building is essential for any effective counseling outcome. In this phase, a good counselor attempts to establish a climate of acceptance, warmth, support, openness, and mutuality. He/she does this by listening to the employees' problems and feeling, by communicating his/her understanding to the employees, and by expressing a genuineness of interest in them.

ii. **Exploration:** In the exploration phase, besides accepting the employees, listening to them, and establishing a climate of openness, the counselor attempts to understand as well as help the employees understand their own situational strengths, weaknesses, problems, and needs. Counseling skills lie in this.

iii. **Action planning:** In the action planning stage, the counselor and the employee jointly workout or plan specific action steps for the development of the employee.

7. Performance Achievement

(a) **To establish rapport or prepare a base for communication**: This step is very necessary as it generates the necessary confidence in the subordinate and assures him of his superior's genuine interest in helping him

(b) **To explore information about performance:** During this stage the counselor puts to the subordinates various questions, which may elicit elaborate information on the latter's achievement, strengths, failures and short coming. The object is to make the subordinate himself to introspect and define his strengths and weakness.

(c) **To help define the future goal and internalize the problem:** Once the counselor has succeeded in making the subordinate reflect on his strengths and weaknesses he can easily make him realize what his problems are. There is generally a tendency to attribute ones' failures, weaknesses or shortcoming to external reasons. The counselor has to help the subordinate to internalize the problem and its causes. He must also help him in setting his future goals.

(d) **To draw action plans:** Both the counselor and the subordinate jointly consider all possible alternative solutions to the problems. Their pros and cons are weighed and the best alternative selected and a stepwise action plan is prepared along with the timetable. The subordinate is allowed to monitor the plan himself. The plan is reviewed by the two parties at regular intervals.

The various requisites of the counseling are listed below:

i. Individual's Desire to Improve
ii. Continuous Dialogue
iii. Interest of Superior in Subordinates' Development
iv. Mutual Participation

8. Functions

The basic objective of counseling is to help employees to improve their mental health and develop self-confidence, understanding, self-control and ability to work effectively. This objective can be achieved by performing various counseling functions. Functions are:

1. Advice
2. Reassurance
3. Communication
4. Release of Emotional Tension
5. Clarified Thinking
6. Reorientation

9. Qualities of a Counsellor

i. Counsellor should be a morale booster for the employee
ii. Counsellor should be unbiased
iii. Counsellor should have warm manners and social etiquettes
iv. He/she must be well versed in excellent communication skills
v. The counsellor should possess professional qualifications.
vi. Effective listening skill
vii. Pleasing personality
viii. Immense Patience
ix. Compassionate approach
x. Non-judgmental towards a problem or an employee
xi. Research-Oriented
xii. Empathetic and sympathetic towards the employees
xiii. Discrete outlook
xiv. Come forward for redressed

10. Factors for Successful Counseling Program

The corporate world is getting changed and so the HR process also. Counseling not only helps employees but also show how much the organization cares for employees.

Employee Counseling needs to be tackled carefully, both on the part of the organization and the counselor.

i. The counsellor should be either a professional or an experienced, mature employee.
ii. The counsellor should be flexible in his/her approach and a patient listener.
iii. The counsellor should have the warmth required to win the trust of the employee
iv. The counsellor be active and effective listener

v. Time should not be a constraint in the process.
vi. The counsellor should identify the morale and spirit of the employee
vii. Counsellor must give workable advice

11. Benefits of Counselling

i. It helps employees to tackle with the problems effectively
ii. Employees are able to sort out their problems with ease
iii. Counselling helps in taking correct decisions
iv. Counselling new perspective and positive outlook
v. It may reduce absenteeism of employee
vi. It may prevent termination
vii. It reduces the cost of hiring
viii. It results in smooth coordination
ix. It helps the individual to understand and help him/herself
x. It provides alternate solutions to problems
xi. Counselling helps in Coping with the situation
xii. Removes frustrations and stress through Counselling

H. Mentor System

Definition

Mentoring is a process where a more experienced or more knowledgeable person helps to guide a less experienced or less knowledgeable person. The goal of mentoring is to develop the skills and knowledge of the less experienced person in order to help them reach their potential.

Mentoring is a system of semi-structured guidance whereby one person shares their knowledge, skills and experience to assist others to progress in their own lives and careers. Mentors need to be readily accessible and prepared to offer help as the need arises - within agreed bounds.

Purpose of a Mentoring Program

i. To develop leadership capability
ii. To increase the skill levels of employees
iii. To increase employee engagement

Definition of Mentorship

ATD (Association for Talent Development) "Mentoring is a reciprocal and collaborative at-will relationship that most often occurs between a senior and junior employee for the purpose of the mentee's growth, learning, and career development."

Five Stages of Mentorship

Every mentoring bond undergoes a series of evolutionary stages, often termed as the mentoring cycle. David Clutterbuck, author and thought leader on the subject of coaching and mentoring identifies these transitions as: building rapport, setting direction, making progress, winding down and finally, moving on.

Stage 1 - Building rapport

The mentoring journey begins with the mentor and mentee gauging compatibility in an initial assessment. Key elements, as highlighted by Megginson and colleagues in "Mentoring in Action" (2006), include:

i. Shared values and alignment.
ii. Mutual respect.
iii. Clarity on the relationship's objectives.
iv. Consensus on expected roles and behaviours.

A strong rapport forms the bedrock of a successful mentoring bond.

Stage 2 - Setting direction

This phase is about goal-setting, defining the relationship', and understanding its immediate and future course. Often, this stage in the mentor program, along with the rapport building, can be established in the initial meetings.

Stage 3 - Making progress

The most intense learning occurs in this phase of the mentor program for both participants. Efficient mentors will:

i. Create a structured yet comfortable environment.
ii. Agree on the meeting's objective.
iii. Understand issues from the mentee's viewpoint.
iv. Promote analytical thinking and draw from personal experiences.
v. Boost the mentee's confidence.
vi. Discuss various possibilities.
vii. Recap and agree on mutual actions.
viii. Propose topics for the next meet.

Stage 4 - Winding down

This phase involves reflection, where the mentor and mentee review their journey and celebrate their achievements.

Stage 5 - Moving on

The formal mentor-mentee bond concludes, transitioning to a more casual relationship, possibly as friends or professional peers.

The 4 C's of Mentorship

<table>
<tr><td>1. Conversation</td><td rowspan="2">The 4 C's of mentorship</td><td>2. Connection</td></tr>
<tr><td>3. Community</td><td>4. Culture</td></tr>
</table>

i. **Conversation**: Quality mentoring promotes meaningful dialogue. Beyond standard workplace interactions, these conversations entail deep listening, mutual problem-solving, and shared goals.

ii. **Connection:** Workplace engagement often hinges on meaningful relationships. Through mentorship programs in the workplace, unlikely connections are forged, broadening organizational understanding.

iii. **Community**: Belonging is key to employee engagement. Through mentoring, individuals feel more connected to their workplaces.

iv. **Culture**: A mentoring culture thrives on rich dialogue, diverse connections, and a united community. It reflects an organization's commitment to growth and development.

What is a Mentoring Program?

Mentorship programs are a structured and organized initiative that facilitates the pairing of experienced individuals (mentors) with less experienced ones (mentees) to provide guidance, support, and knowledge transfer.

Key Components of a Mentoring Program Include

i. **Matching process:** This is where mentors and mentees are paired together. Some programs have a formal mentor matching process managed by an administrator, while others might be more informal, allowing mentees and mentors to self-select each other.

ii. **Training**: Many programs offer training sessions for both mentors and mentees to equip them with the necessary skills and set clear expectations for the mentorship relationship.

iii. **Feedback and evaluation**: Regular check-ins, feedback sessions, and evaluations help assess the effectiveness of the mentorship and make necessary adjustments.

iv. **Goals and objectives**: Setting clear objectives at the outset helps guide the mentorship process and provides a framework for the mentee's growth and development.

v. **Duration**: Mentorship programs can vary in length, from short-term initiatives (like a few months) to long-term partnerships that last several years.

vi. **Support structures**: This might include resources, regular meetings, workshops, and other support mechanisms to facilitate the mentorship process.

vii. **Peer networking**: Providing opportunities for colleagues to connect through events and peer-to-peer mentoring is a great way to support your core mentorship program, encouraging social learning.

Mentors and mentees with these qualities will easily build a relationship that is mutually beneficial.

i. Important factors to evaluate mentoring program
ii. Engagement levels of participants their enthusiastic attitude
iii. Goal achievement setting of goal by both for achievement
iv. Qualitative feedback - how do they describe the program to others?

A mentoring program doesn't start and end at pairing. To ensure a successful workplace mentoring program administrators should keep their finger on the pulse of all pairings and make adjustments as needed.

Purpose of a Mentoring Program

The main purposes are:

i. To develop leadership capability:
ii. To increase the skill levels of employees
iii. To increase employee engagement
iv. Smaller companies place a high emphasis on mentoring to increase employee skill levels (75 per cent).
v. Larger companies are more inclined towards mentoring to support diversity, equity, and inclusion, with 48 per cent prioritizing this compared to 40 per cent of mid-size and just 21 per cent of small organizations.

Qualities of Good Mentees and Mentors Include

1. Drive to succeed
2. Open to learning and new perspectives
3. Shows initiative
4. A positive attitude
5. Leadership skills or capabilities
6. Good time management skills
7. Clear communication
8. Respect to Humanity

Mentors and mentees with these qualities will easily build a relationship that is mutually beneficial.

Benefits of Mentor, Mentee and Organization

Benefits to Mentees	*Benefits for Organizations*
1. Learn the workplace culture	1. Attract the right talent
2. Enhance skill development	2. Help advance careers
3. Networking opportunities	3. Retain high performers
4. Get promoted more	4. Increase employee engagement

The 9 Types of Mentoring Models

From traditional one-on-one guidance to innovative hybrid models, workplace mentoring programs offer different avenues to achieve organizational goals while supporting mentors and ensuring mentees feel engaged and advance their careers.

i. **One-on-one mentoring**: It is a classic mentoring programs, this classic approach involves an experienced mentor guiding a singular mentee.

ii. **Peer Mentoring**: Two individuals of similar professional standing or age join forces. By sharing knowledge and experiences, they aim to grow together and maintain mutual accountability.

iii. **Group mentoring**: One mentor engages with multiple mentees simultaneously. By doing this, a mentor can reach more individuals in a shorter time span.

iv. **Reverse mentoring**: This model flips the traditional setup. A younger or less experienced individual mentors someone senior or more experienced. This acknowledges that learning can flow in both directions.

v. **Flash mentoring**: These are short, focused mentoring sessions designed for quick knowledge transfer or skill acquisition.

vi. **Team mentoring**: One mentor for many mentees, team mentoring involves several mentors guiding a group. It's beneficial when a group of individuals is working towards a common objective. This setup fosters a diverse learning environment as multiple mentors bring varied perspectives and knowledge.

vii. **Virtual mentoring**: With the rise of remote work, virtual mentoring has gained prominence. Different mentorship styles can be executed remotely using various digital tools

viii. **Mosaic mentoring**: An emerging, hybrid style where mentees collaborate with multiple mentors or explore various mentoring program frameworks.

ix. **Functional mentoring**: A more project-focused approach, where the mentoring revolves around a clear, predefined goal.

The 4 Types of Mentors

i. **Peer mentors:** Peer mentors can provide guidance, advice, and support to their peers. They are usually from similar backgrounds or groups of similar age and experience so that they can identify with each other's struggles and goals.

ii. **Career mentors:** Career mentors are those who have developed their professional experience and can provide guidance to help others achieve their career goals. They typically have advanced knowledge in the field.

iii. **Reverse career mentors:** Reverse career mentors are less experienced employees who mentor a senior colleague. Reverse mentoring bridges the generational gap and helps build workplace connections with senior management.

iv. **Life mentors:** Life mentors are individuals who provide guidance and advice to those seeking it for their life skills, confidence, knowledge, and expertise in a given area.

Qualities of Good Mentor

i. Experience in the same field as their mentee

ii. Good communication skills

iii. A willingness to listen and support others, regardless of experience level

iv. An understanding of different perspectives and cultures within the workplace

v. The ability to help their mentees develop new skills or refine existing ones

vi. Emotional intelligence

vii. Ability to provide constructive feedback

viii. A positive outlook and a love for their profession

Four Qualities of a Great Mentor

1. **Patience**: Understand their mentee's challenges, motivations and goals.
2. **Empathy**: Understand source specific issues
3. **Knowledgeable**: Wealth of knowledge on topics
4. **Open-mindedness**: Think outside the box to help mentee
5. **Communication**: Excellent communication skills
6. **Integrity**: Highest level of integrity

I. Self-Renewal System

Self-renewal system: It is not enough to develop individuals and teams in the organizations but occasionally there is a need to renew and re-juvenate the organization itself. Following are some of the sub systems that can be utilized to renew the organization.

J. Survey

A survey is a method of gathering information using relevant questions from a sample of people with the aim of understanding populations as a whole. Surveys provide a critical source of data and insights for everyone engaged in the information economy, from businesses to media, to government and academic.

Types of Surveys

There are four modes of surveys that are commonly used.

i. Face-to-face surveys
ii. Telephone surveys
iii. Self-administered paper and pencil surveys
iv. Self-administered computer surveys (typically online)

The Benefits of Surveys

i. Responses, faster
ii. They're cheaper to run
iii. Online surveys are more accurate
iv. Reach new audiences

K. Action Research

Definition

Action research is focused on solving a problem or informing individual and community-based knowledge in a way that impacts teaching, learning, and other related processes. It is less focused on contributing theoretical input, instead producing actionable input

Background

Action research is a research method that aims to simultaneously investigate and solve an issue. In other words, as its name suggests, action research conducts and takes action at the same time. It was first coined as a term in 1944 by MIT Professor Kurt Lewin. A highly interactive method, action research is often used in the social sciences, particularly in educational settings.

Types of Action Research

There are two common types of action research: participatory action research and practical action research.

i. **Participatory action research** emphasizes that participants should be members of the community being studied, empowering those directly affected by outcomes of said research. In this method, participants are effectively co-researchers, with their lived experiences considered formative to the research process.

ii. **Practical action research** focuses more on how research is conducted and is designed to address and solve specific issues.

Both types of action research are more focused on increasing the capacity and ability of future practitioners than contributing to a theoretical body of knowledge.

Action Research Models

Action research is often reflected in 3 action research models: Operational (sometimes called technical), Collaboration and Critical reflection.

i. **Operational (or technical) action research** is usually visualized like a spiral following a series of steps, such as "planning → acting → observing → reflecting."

ii. **Collaboration action research** is more community-based, focused on building a network of similar individuals (*e.g.*, college professors in a given geographic area) and compiling learning from iterated feedback cycles.

iii. **Critical reflection action research** serves to contextualize systemic processes that are already ongoing (*e.g.*, working retroactively to analyse existing school systems by questioning why certain practices were put into place and developed the way they did).

Action Research vs. Traditional Research

Action research differs sharply from other types of research in that it seeks to produce actionable processes over the course of the research rather than contributing to existing knowledge or drawing conclusions from datasets. In this way, action research is **formative**, not **summative,** and is conducted in an ongoing, iterative way.

Point of Difference	*Action research*	*Traditional research*
Purpose	1. Solve immediate problems 2. Improve existing systems	1. Draw conclusions from data 2. Advance existing knowledge 3. Provide generalized and reliable findings
Context	1. Reactive, derived from surroundings 2. Usually not theoretical in nature	1. Focused on crafting strong hypotheses and seeking cause relationships between variables 2. Derived from theory
Significance	1. Practical	1. Statistical

Advantages and Disadvantages of Action Research

Advantages	*Disadvantages*
Action research is highly adaptable, allowing researchers to change their analysis to their individual needs and implement practical individual-level changes.	Due to their flexibility, action research studies are plagued by very limited generalizability and are very difficult to replicate. They are often not considered theoretically rigorous due to the power the researcher holds in drawing conclusions.
Action research provides an immediate and actionable path forward for solving entrenched issues, rather than suggesting complicated, longer-term solutions rooted in complex data.	Action research can be complicated to structure in an ethical manner. Participants may feel pressured to participate or to participate in a certain way.
Done correctly, action research can be very empowering, informing social change and allowing participants to effect that change in ways meaningful to their communities.	Action research is at **high risk for research biases** such as selection basis social Desirability basis, or other types of cognitive biases.

Stages of Action Research

Stage 1 – Plan: The first step is to develop an action plan to guide the research process. The research design aims to address the study's question. The research strategy outlines what to undertake, when, and how.

Stage 2 – Act: The next step is implementing the plan and gathering data. At this point, the researcher must select how to collect and organize research data. The researcher also needs to examine all tools and equipment before collecting data to ensure they are relevant, valid, and comprehensive.

Stage 3 – Observe: Data observation is vital to any investigation. The action researcher needs to review the project's goals and expectations before data observation. This is the final step before drawing conclusions and taking action.

Stage 4 – Reflect: This step involves applying a prospective solution and observing the results. It's essential to see if the possible solution found through research can really solve the problem being studied.

The researcher must explore alternative ideas when the action research project's solutions fail to solve the problem.

The Steps to Conducting Action Research

Action research is a systematic approach, researcher, educators, and practitioners use to identify and address problems or challenges within a specific context. It involves a cyclical process of planning, implementing, reflecting, and adjusting actions based on the data collected.

The steps to conducting action research:

i. Review existing knowledge
ii. Plan the research
iii. Collect data
iv. Analyse the data
v. Reflect on the findings
vi. Develop an action plan
vii. Implement the action plan
viii. Evaluate and monitor progress
ix. Reflect and recommend

11.5 Organisational Development Interventions and Culture System (Vision, Mission and Goals, Values of Organisation, Communication, Get together and Celebrate in HR, Task Force in HR, Small Group in HR)

A. Organisational Development Interventions

Organizational development (OD) interventions are programs and processes designed to improve an organization's effectiveness by creating actions that change leadership styles, organizational culture, structures, or behavioural patterns.

Meaning and Definition

An OD intervention is defined as, "the set of structured activities in which selected organisational units (target groups or individuals) engage with the task or a sequence of task where the task goals are related directly or indirectly to organizational improvements According to Rober Zawacki, "OD intervention is a sequence of activities actions and events intended to help an organisation improve its performance and effectiveness".

Importance of OD Interventions

OD interventions are very important to the organisations as discussed below:

1. Boosts the quality of employee performance
2. Leads to perpetually improving organisation
3. Enhances the quality of products/services
4. Enhances the Competitive Edge of the organisation
5. Increases the profitability of the organisation

Types of OD Interventions

Organisational development is very complex, and the interventions are equally as intricate. OD interventions can be categorized into four groups:

1. Human process interventions
2. Techno-structural interventions
3. Human resource management interventions
4. Human progress interventions

1. *Human process interventions*

Human process interventions are organisation development interventions related to interpersonal relations, group, and organizational dynamics. These were the earliest form of interventions and are often aimed at improving communication within the workplace.

2. *Techno-structural interventions*

Techno-structural interventions are targeted toward structural and technological issues such as organisational design; work redesign, and employee engagement. Techno structural interventions are becoming increasingly relevant to today's rapidly changing markets and technological landscape. Some important interventions are as follows:

i. **Organizational (structural) design:** This change process concerns the organisation's division of labour - how to specialize task performances. Diagnostic guidelines exist to determine which structure is appropriate for particular organizational environments, technologies, and conditions.

ii. **Total quality management:** Total Quality Management (TQM) is a management framework based on the belief that an organization can build long-term success by having all its members, from low-level workers to its highest ranking executives, focus on improving quality and thus delivering customer satisfaction.

iii. **Work design:** Designing work in a way that leads to optimum productivity is called work design. Work design is the division of work tasks assigned to an individual in an organization that specifies what the worker does, how, and why. Effective design contributes to the achievement of organizational objectives, motivation, and employee satisfaction.

iv. **Job enrichment:** Job enrichment is part of work design. The goal here is to create a job that is interesting and challenging for the person doing it. Job enrichment is a process that is characterised by adding dimensions to existing jobs to make them more motivating.

3. *Human resource management interventions*

Human resource management interventions impact areas such as performance management, talent development and wellbeing in the workplace. Strategic change interventions: Strategic change interventions revolve around transformational change, restructuring, and uniting two or more organizations together during a merger. These interventions can be of following types:

i. **Performance management:** Organizations use performance management (reviews) to support employee training, career development, compensation decisions, and promotions. Performance management is a corporate management tool that helps managers monitor and evaluate employees' work.

ii. **Developing talent:** Organisations need strong talent management practices to stay relevant and competitive in the changing landscape. Some methods for developing talent are as follows:

   1. Individualised career planning
   2. Internal or external coaching
   3. Task/job rotations.
   4. Mentorship programs
   5. Internal or external workshop
   6. Conferences
   7. On-the-job training
   8. Leadership training

iii. **Wellness interventions**: A wellness intervention is a combination of program elements or strategies designed to produce behaviour changes or improve health status among individuals or an entire population.

iv. **Reward systems:** Reward systems in organizations can be thought of as serving three distinct but related purposes: attracting, motivating, and retaining employees.

4. *Human progress interventions*

Some of the important human progress interventions are as follows:

i. **T-groups sensitivity training is also known as T-groups (T for training):** This technique is based on the assumption that a manager's behaviour is not how he thinks, how others view his behaviour. The basic objective of sensitivity training is to change the behaviour of the people through unstructured group interactions. With this training a person understands how his behaviour affects others and his reaction to the behaviour of others.

   T-Groups have the following characteristic features:

   1. Members (10 to 15 individuals) is brought together in a free and open environment, away from work places.
   2. No formal agenda is provided for the meeting.

3. The participants discuss freely aided by a facilitator. The role of facilitator is to call attention from time to the ongoing process within the group.
4. The procedure tends to develop introspection and self-examination, individual personality, and group interaction. Processes and relationships become the focus of discussion. The role of the facilitator is to create an opportunity for the members to express their ideas, beliefs and attitudes.

ii. **Process consultation:** According to Edgar Schein, "Process consultation includes a set of activities on the part of a consultant which helps the client to perceive, understand and act upon process events which occur in the client's environment." The assumptions underlying the process consultation model are:

1. Managers often need special diagnostic help in knowing what is wrong with the organisation.
2. Most managers have constant desire to increase organisational effectiveness, but they need help in deciding how to achieve it.
3. Managers can be effective if they learn to diagnose their own strengths and weaknesses without exhaustive and time consuming study of the organisation.
4. The outside consultant cannot learn enough about the culture of the organisation, to suggest new reliable courses of action. He should, therefore, work jointly with the members of the organisation.
5. The client must learn to see the problem for himself, understand the problem and suggest a remedy. The consultant should provide new and challenging alternatives for the client to consider. However, the decision making authority on these alternatives about organisational changes remain with the client.
6. It is essential that the process consultant is an expert in diagnosing and establishing effective helping relationships with the client.

iii. **Team development:** Team development is a process of diagnosing and improving the effectiveness of a work group with particular attention to work procedures and interpersonal relationships within it, especially the role of the leader in relation to other group members.

There are three approaches to team development explained as follows:

1. The consultant will interview members of the team individually to know their feelings, attitudes and perceptions of team effectiveness. After that, the consultant will arrange a meeting of group away from the organisation and provide them feedback data which will be discussed in detail. The set of priorities will be worked out and action plan will be formulated for resolving the problem.

2. In this approach, each team member will discuss with another member his perceived roles and also team feeling, so as to make more meaningful and productive contribution.
3. The consultant will regularly attend the team meetings. He will observe how the team is accomplishing the group tasks and maintaining roles.

iv. **Third party peace making:** Third party peace-making as the name suggests, focuses on the interventions by a third party to resolve the confecting, Situations. It is aimed at the analysis of involved processes, diagnose the conflict's causes and with the assistance of a third party consultant resolves the conflict effectively. These techniques may be:

1. Wisely choosing the place
2. Selecting the proper environment
3. Using effective intervention strategies
4. Setting an appropriate agenda for the meeting
5. Helping the parties in conflict to own up to their problems

B. Culture System

1. Vision

Provides a mental image of an extraordinary accomplishment the organization would like to achieve. Articulates the overreaching, long-term goals of the enterprise. Looks beyond the present to see what the future could be.

Seven characteristics of a vision statement:

i. Provides a mental image of an extraordinary
ii. Articulates the overreaching, long-term goals of the enterprise.
iii. Looks beyond the present to see what the future could be.
iv. Asserts what the organization can be at its best.
v. Presents *a target that beckons.*
vi. Reflects, often times, what is unique about organization.
vii. Is vivid, motivating, and memorable

Elements of organizational vision: Best practice is to provide team members with purpose and direction. Successful leaders set organizational vision by:

i. Creating a vision
ii. Developing a mission
iii. Executing with values

2. Mission and Goals

The goals of organizational development help define a company's purpose, assist its business growth and achieve its financial objectives. Setting specific structural goals can also facilitate an organization to attain progress and complete the tasks to meet those business goals.

Difference between a vision and a mission statement:

(i) A vision statement is short
(ii) A vision statement generally contains a super ordinate goal
(iii) A mission statement often includes nature and direction of the business
(iv) A vision statement is the bridge between present and future

Mission of an Organization

A mission statement defines the organization's business, its objectives, and how it will reach these objectives. An organizational mission is a statement directed both internally and externally, so it helps inform how stakeholders at all levels perceive the organization.

Goals of Organization

Organizational goals are measurable objectives that are set to help an organization attain aims like higher profit, enhanced customer satisfaction, larger market share, employee retention or innovation in the goods and services that it provides

The four goals of organizational behaviour are to describe, understand, predict and control.

Types of Organizational Goals

Goal	*Interpretations*
1. Strategic Goals	These are long-term objectives that define the overall direction and vision of the organization.
2. Operational Goals	Operational goals are shorter-term objectives that are more specific and measurable. They focus on day-to-day activities for improving efficiency, reducing costs, or increasing productivity.
3. Financial Goals	Financial goals relate to the financial performance and stability of the organization.
4. Market Goals	Market goals pertain performance of markets, or enhancing the organization's reputation and brand recognition.
5. Customer Goals	Customer goals are cantered around meeting the needs and expectations of customers.
6. Employee Goals	Employee goals focus on the development and well-being of the organization's workforce.
7. Innovation Goals	Innovation goals emphasize the importance of creativity and continuous improvement.

Well-defined goals can help organizations to:

i. Improve Performance and Productivity.
ii. Increase Employee Engagement and Motivation
iii. Make Better Decisions
iv. Allocate Resources More Effectively.
v. Gain a Competitive Advantage

Six Ways to Achieve Organizational Goals and Objectives

I. Set SMART (Specific, Measurable, Achievable, Relevant and Time-bound) Goals

SMART goals are specific, measurable, achievable, relevant and time-bound. Setting SMART goals helps the companies to ensure that goals are well-Specific: Be specific as possible defining goals. A specific goal answers questions like:

i. What exactly to be achieve?
ii. Who's the responsible person/team for it?
iii. What steps needed to be taken to achieve it?

These questions help to get to the heart of the company's aim.

Measurable: Quantifying the company's goals makes it easier to track progress and know when we've reached the finish line. For example, instead of saying "Increase sales," say "Increase sales by 8 per cent in the next quarter."

Achievable: Goals should be realistic. Does the organization have the resources and capabilities to achieve the desired goal? Have other organizations done it successfully before?

Realistic: A SMART goal should be realistic. It can be realistically achieved, given the available resources and time.

Time-Bound: A SMART goal must be time-bound because it has a start and finish date. If the goal is not constrained by time, there will be no urgency and, therefore, the motivation will be less to achieve the goal.

II. Clear Plan of Action

Once set SMART goals, there is need to develop a clear plan of action to achieve them. This plan should identify the steps that need to be taken, the resources that will be needed, and the timeline for completion. It is important to involve key stakeholders in the planning process to ensure everyone is aligned on the goals and how they will be achieved.

III. Mitigate Distractions

In today's fast-paced world, staying focused and avoiding distractions can be difficult. However, it is essential to eliminate distractions as much as possible when working towards important goals.

Tips for mitigating distractions:

i. Set aside time to work on goals without distractions
ii. Take breaks regularly
iii. Use productivity tools
iv. Employee efficient time management

IV. Time Management

Time management is essential for achieving organizational goals.

i. Set Deadlines
ii. Break down tasks in to smaller ones
iii. Avoid multitasking
iv. Take breaks

The "Eat That Frog" technique is a time management strategy that encourages tackling the most difficult task first thing in the day. This will help to get the most challenging task out of the way and free up our time and energy for other tasks.

Here are some tips for using the "Eat That Frog" technique:

i. Identify frog
ii. Tackle frog first thing in the morning
iii. Break down frog into smaller tasks.
iv. Set deadlines

V. Apply the Pareto Principle

The Pareto principle, also known as the 80/20 rule, states that 80 per cent of outputs come from 20 per cent of inputs. This means we can achieve significant results by focusing on the most important 20 per cent of our tasks.

VI. Always Track Progress

It is important to track progress toward goals. Here are some tips for tracking progress:

i. Set milestones
ii. Track progress regularly
iii. Celebrate Successes

3. Values of Organization

Definition

Organisational values are a set of core beliefs held by an organisation. They act as guiding principles that provide an organisation with purpose and direction and set the tone for its interactions with its customers, employees and other stakeholders.

Buffer: Six Core Values	*Kellogg's: Six Core Values*
1. Default to transparency 2. Cultivate positivity 3. Show gratitude 4. Practice reflection 5. Improve consistently 6. Act beyond yourself	1. We act with integrity and show respect 2. We are all accountable 3. We are passionate about our business, our brands and our food 4. We have the humility

Purpose of Organisational Values

Almost every business leader has certain unwritten values and principles that they expect their employees to embody.

Benefits of formalising organisational values are,

I. **Sets apart from competitors:** Organisational values are, by definition, unique to each organisation. They show the public what makes different and can even give a competitive advantage.

II. **Guides employee decision-making:** By clearly defining organisational values, one can ensure every decision an employee makes is in line with the organisation's goals and mission.

III. **Attracts like-minded talent:** The organisational values can help to attract candidates who share the same values and fit the culture of organisation.

IV. **Attracts customers:** Today's customers like to see that the brands they buy from have values that align with theirs

V. **Improves employee engagement:** The organisational values employees something to believe in and work towards. This makes them feel more connected to organisation, improving engagement and motivation.

VI. **Informs strategic direction:** Just as organisational values should guide any decisions employees make, they also underpin the major strategic decisions make about organisation's future.

Developing Organisational Values

Steps

i. Put a Team Together

ii. Brainstorm our Values

To guide brainstorming sessions we may have following questions.

i. What is the core purpose of the organisation?
ii. What are our personal values?
iii. How do we want the world to see our organisation?
iv. What behaviours do we want to encourage employees?
v. Which values are already embedded in organisation's structure, culture and processes?

4. Communication

Definition

HR communication is a two-way process that involves information being sent from HR to employees and from employees up to HR. Employees who feel their voice is heard are 4.6 times more likely to be empowered to perform their best work; 5 per cent increase in employee engagement can lead to a 3 per cent jump in revenue.

Effective communication can increase productivity while preventing misunderstandings. Leaders who can explain the benefits of HR plans, for example, are more likely to cultivate employee buy-in.

There are four forms of communication *i.e.*, verbal, non-verbal, visual and written communication

HR Communication

Communication is the foundation of all human relationships. In the workplace, it facilitates the exchange of information and knowledge. It's essential for individual and team success and enables workers to develop strong relationships and establish themselves as collaborative team members who want to help the organization reach its objectives.

Importance of HR Communication

The key importance points are,

i. Improved engagement
ii. Increased productivity
iii. Increased retention

The 4 Types of Organizational Communication

Information flows in 4 directions: downward, upward, horizontally, and diagonally. The direction is usually determined by the size, nature, and structure of the organization.

I. **Upward communication:** Employees communicate with upper management to provide feedback, complaints, or suggestions. Upward communication is becoming increasingly popular because it encourages a participatory work environment.

II. **Downward communication:** In downward communication, information and messages flow top-down through the organization's hierarchical structure.

III. **Diagonal communication:** Diagonal communication occurs between departments with people at different levels of the organization. This is a dialogue between employees of differing ranks and functions who are not in the same chain of command.

IV. **Horizontal communication:** This is when people of the same level in an organization communicate, usually to coordinate work between departments. This flow of communication is used to run departments when it's essential to gather information across members of the same work group or same level in the organization.

Developing a Comprehensive HR Communication Strategy

I. **Define the employee segments:** Employee segmentation enables to identify different employee groups in the organization based on similarities such as roles, geographical area, gender, age, behaviours, and interests so on we can design customized practices and policies based on the common characteristics and needs of each group to increase the relevance of communication.

II. **Value proposition and communication goals:** Defining the value proposition and communication goals starts with determining the full mix of benefits that it promises to deliver to all our internal customers. This requires to think about the organizational strategy and the role of internal communication in helping to achieve that.

Types of HR Communication Value Propositions

i. **Performance:** Improving HR communication so that it's faster, easily accessible

ii. **Customization:** Tailoring communication services to the specific needs of internal customer segments

iii. **Getting the job done:** Creating value by enabling employees to get the job done better and more efficiently.

iv. **Convenience:** Making things more convenient or easier to use can create substantial value for employees.

III. **Define HR communication channels:** Today, digital communication channels email, internal communication platforms, employee collaboration software, intranets, internal podcasts, internal company blogs are the most popular in the workplace.

IV. **Determine internal customer relationships:** What type of relationship does each of our internal customer segments expect from HR? We need to clarify the type of relationship we want to have with each employee segment.

V. **Define HR communication activities:** Consider the best activities to deliver important messages to internal customer. Key activities should focus on fulfilling the value proposition, reaching internal customer segments and maintaining internal customer relationships.

VI. **Determine the resources required:** Determine what resources we require to best deliver message. Engaging customers is a strength of marketing, so the marketing team may be able to support in creating engagement programs in the workplace.

VII. **Determine the communication costs:** Determine the costs associated with HR communication strategy and create a budget. A properly defined internal communications budget helps turn communication flowing across the business into valuable and measurable business assets.

VIII **Define the HR communication value drivers:** Organizations rely on HR professionals to relay important information and policy changes to employees. Communicating effectively not only ensures that everyone is kept informed, but it also can help the business to remain compliant and avoid unnecessary challenges. Perhaps even more importantly, it leads to increased employee trust, engagement, and satisfaction.

Internal Communication Metrics

I. **Employee engagement rates:** The company intranet can help to measure employee engagement with key performance indicators (KPIs) such as readership, participation and social metrics like the number of likes, shares, and comments a piece of content generates.

II. **Open rates or Intranet read receipts:** Read receipts on a company intranet measure whether an employee actually acknowledged receiving and reading important updates.

III. **Page visits and logins:** Capture unique page views and other observable metrics for day-to-day optimization.

IV. **Adoption rates for new apps:** When introducing a new employee app, tracking the adoption rate is critical.

V. **Employee feedback:** Encourage employees to ask questions and give their opinions about what is happening within the company.

VI. **Employee turnover rates:** Employees who feel engaged at work are not likely to leave their job. For this reason, an important internal communication KPI to measure is employee turnover rate.

5. Get Together and Celebrate in HR

Ten special ways to celebrate Human Resource professional day.

I. **HR employee spotlights:** Sending a mass email saying "thanks, HR team!" might be easy, but great recognition is more personal. Instead, consider writing up employee spotlights for each member of HR team.

II. **Take HR team out:** Offering a team building day out can be a great way to celebrate human resources department.

III. **Build a HR appreciation wall:** If employees mostly work in one office, this celebration of gratitude is fun and easy: print out pictures of HR team and give employees some information about what interesting and experienced people they are. This "wall of fame" gives HR employees more visibility and humanity within company while making them feel great too.

IV. **Cater a lovely meal:** Providing a delicious catered breakfast or lunch is a great way to reward HR professionals for their hard work. Of course, this is easiest if everyone is in the office together, but with the rise of remote work, that's not always the case.

V. **Give shootouts on social media:** Recognizing HR team internally is great but so is a little external validation. Post on social media feeds about how great HR team is and what they've accomplished over the past year to spread the love.

VI. **Make a HR appreciation video:** Videos don't need to be polished and expensive to be fun in fact, if they're too polished they can feel inauthentic. Taking the time to make a quick appreciation video, from leaders, HR managers, or employees with testimonials about how much their work is appreciated, can go a long way in helping HR team truly feel how much their work is seen and valued by your company.

VII. **Send thank you notes from leadership:** Asking a few leaders, especially head of HR, to write notes of appreciation to HR team members is a lovely way to celebrate the day. Plus, it doesn't cost anything!

VIII. **Send an HR care package:** This celebration is great for fully remote teams sends each HR team member a care package for the day.

IX. **Offer HR employees support:** Working in HR can be a challenging job, especially when teams are dealing with issues like layoffs or hiring freezes. The HR team is tending to employee well-being for everyone else, and so their needs might be neglected.

X. **Hold an HR knowledge-sharing session:** HR seminar can be arranged to share knowledge and experiences.

6. Task Force in HR

Task force is a management technique that produces quality decisions and employee commitment.

Definition

The **Task Force** is a type of a group, formed temporary, in which people from different disciplinary backgrounds come together to perform a specific task or mission. These are different from the committees in the sense, these are temporary and has broader powers of action and decision, greater responsibilities for investigation, analysis, planning and research.

What is a Task Force?

A Task Force is an interim group that is assembled to complete a certain task or address a specific issue. Members of the task force are typically experts in different fields from different departments or organizations. Task forces are goal-oriented teams that tackle challenging issues that call for collaboration and disband when their assignment is over. To ensure efficient oversight and results, they report to a higher authority and adhere to a set aim, legitimate power, and financial resources.

Features of Task Force

I. **Goal-oriented:** This indicates that a task force has been established with a specific purpose, such as conducting research, developing new regulations, or responding to a crisis.

II. **Transient:** Unlike permanent teams or departments, task forces are usually disbanded after they have effectively completed their job.

III. **Cross-functional:** Regarding cross-functionality, it should be noted that work groups frequently include individuals from various professional fields who make use of their unique competencies and perspectives.

IV. **Authority and resources:** Typically, task force participants receive particular authority and resources necessary to meet the goals they have established.

V. **Reporting and accountability:** Task forces typically turn in their reports to a higher authority for oversight and governance. This higher authority is most likely a member of senior management, a government agency, or an organizational leader.

Reasons to Use Task Force

I. **Complex obstacles:** Task Forces are set up to address complex issues that call for specialized knowledge and coordinated efforts from several parties.

II. **Quick response:** Task Forces can be quickly formed to gather resources and respond in an orderly manner in the event of an emergency or crisis, minimizing damage and facilitating a speedy recovery.

III. **Cross-functional collaboration:** They make it possible for individuals with various skills and perspectives to collaborate, which fosters innovation and comprehensive problem-solving.

IV. **Interim focus:** Task Forces provide organizations with concentrated attention to specific objectives, enabling them to combine their resources and efforts efficiently without being constrained by long-term frameworks.

V. **Expertise aggregates:** To identify the most effective tactics and solutions, task forces pool the expertise of specialists in a range of domains.

VI. **Political or organizational mandates:** Task Forces are frequently formed by governments, businesses, and other bodies to carry out certain mandates, address public concerns, or advance organizational objectives.

VII. **Navigating uncertainty:** In situations where there is uncertainty or ambiguity, task forces assist in the analysis, planning, and implementation of strategies to minimize risks and ensure that decisions are made with knowledge.

Problems in Task Force

i. Ambiguity in roles

ii. Obstacles to communication

iii. Group behavior

iv. Limitations on resources

v. Reluctance to adjust

vi. Absence of accountability

vii. Adaptability

7. Small Group in HR

Advantages of small groups include shared decision making, shared resources, synergy, and exposure to diversity. It is within small groups that most of the decisions that guide our country, introduce local laws, and influence our family interactions are made.

Importance of Small Groups to the Organisation - Group Dynamics

I. **Filling in gaps in manager's abilities:** Informal organization may act to fill in gaps in a manager's abilities. If a manager is weak in planning, one of his subordinate may help him informally in such a situation. Informal organization helps in solving the work problems of members. It allows them sharing job knowledge and taking decisions which affect a number of jobs.

II. **Better coordination:** Informal groups evolve short-cuts and eliminate Red-tapism. They facilitate smooth flow of information and quick decision making. All these ensure better coordination among various individuals and departments.

III. **Channel communication:** Informal groups act to fill up the communication gaps which arise in the organization.

IV. **Restrain managers:** Informal groups do not allow the managers to across their limits. They restrict them from acquiring unlimited power and from using their power injudiciously.

V. **Better relations:** A manager can build better relations with his subordinate through informal contacts.

VI. **Norms of behaviour:** Informal groups develop certain norms of behaviour which differentiate between good and bad conduct and between legitimate and illegitimate activities. These bring discipline and order among the employees.

VII. **Satisfied workforce:** Cohesive informal groups provide satisfaction to the workers. As a result, labour turnover and absenteeism are reduced and organization's productivity is increased.

VIII. **Developing future executives:** Informal groups recognize talented workers as their leaders. Such leaders can be picked up by the management to fill vacancies at the junior executive level in the future.

IX. **Shared decision-making:** As group members participate in decision-making, they can raise questions, critically assess ideas, and bring up alternatives or points of view that individual members might not have considered. Thus, groups often make higher quality decisions.

X. **Shared resources:** No single group member has all the experience, skills, time, and talents to help the group progress or reach its goals. Together, they can bring their own resources and access a network of other resources to benefit the group.

XI. **Synergy:** Synergy refers to the idea that the whole is greater than the sum of the parts. In other words, the interactions between group members can create outcomes that are greater than individual members could have accomplished working separately.

XII. **Exposure to New Ideas:** Being part of a small group can widen members' perspectives as they share ideas and listen to each other. Each group member will bring a unique field of experience and background, and as a diversity of opinions, ideas, and viewpoints are shared, every member can learn and grow.

Advantages of Small Groups

i. Advantages of small groups include shared decision making, shared resources, synergy, and exposure to diversity.

ii. It is within small groups that most of the decisions that guide our country, introduce local laws, and influence our family interactions are made.

iii In a democratic society, participation in decision making is a key part of citizenship. Groups also help in making decisions involving judgment calls that have ethical implications or the potential to negatively affect people.

iv. The increased knowledge, diverse perspectives, and access to resources that groups possess relates to another advantage of small groups synergy.

v. Participating in groups can also increase our exposure to diversity and broaden our perspectives.

Benefits

1. Improved communication
2. Faster progress
3. Higher engagement
4. Efficient management
5. Increased productivity
6. Established trust
7. Boosted motivation
8. Maximized resources
9. Better collaboration
10. Equal opportunities
11. More innovation
12. Focused specialization

11.6 Components of HRD System - Performance Appraisal, Potential Appraisal, Task Management System, Development System, Socialisation System in HRM, Governance in HRM, Functions of HRD - Organisational Development, Career Development and Capacity Development

Components of HRD System

It involves a systematic approach to identify the training and development needs of employees. HRD is a process that involves both the employee and the organization working together to achieve mutual goals. It includes a variety of methods such as training, mentoring, coaching, job rotation, and career development.

Components of a Human Resource Management System (HRMS)

i. Application management
ii. Employee management
iii. Organizational management
iv. Benefits management
v. Payroll management
vi. Workplace management
vii. Compensation management
viii. Performance management

A. Performance Appraisal

The term "performance appraisal" refers to the regular review of an employee's job performance and overall contribution to a company. Also known as an annual review, employee appraisal, performance review, or evaluation, a performance appraisal evaluates an employee's skills, achievements, and growth, or lack thereof.

Modern methods of performance appraisal include 360 degree feedback, management by objectives (MBO), psychological appraisals, and the behaviourally anchored rating scale (BARS), to name a few.

In the simplest terms, a performance appraisal is a quarterly or annual evaluation of employee performance and productivity against company goals and objectives. The purpose of the appraisal is to help each employee succeed. The results of the appraisal can help determine promotions and raises or gauge areas where an employee needs to improve their skills.

Five Modern Methods of Performance Appraisal

I. **Management by objectives (MBO):** The management by objectives method is an approach that focuses on improving an organization's performance across the board by articulating clear objectives for the business.

II. **360 degree feedback:** The 360 degree feedback method is designed to get feedback from all the key players an employee impacts in their day-to-day activities.

III. **Behaviourally anchored rating scale (BARS):** In behaviourally anchored rating scale (BARS) method, the employer compares each employee's performance with specific behavioural examples that are anchored to numerical ratings.

IV. **Psychological appraisal:** A psychological appraisal attempts to evaluate how an employee might perform in the future rather than assessing how they have performed in the past.

V. **Assessment canter:** In the assessment centre method, an organization tests its employees on both job capabilities and social interaction skills.

Employee performance appraisals are standard practice in most workplaces and they're critical to helping companies and their employees achieve organizational success and growth.

In the world of HR, there are both traditional and modern methods of performance appraisal. We'll be focusing on the modern methods in this post.

Modern vs. Traditional Methods of Performance Appraisal

Traditionally Method	*Modern Method*
Certain approaches to performance appraisal for example, the ranking method, paired comparison, the grading method, and the critical incidents method, among others are often referred to as traditional methods.	Modern methods of performance appraisal include 360 degree feedback, management by objectives (MBO), psychological appraisals, and the behaviourally anchored rating scale (BARS), to name a few.
Traditional methods tend to measure an employee's past performance and focus on assessing employee personality traits such as initiative, dependability, and leadership potential	Modern methods weigh job achievements more heavily, regardless of the employee's personality traits, which is thought to be a less biased approach.

B. Potential Appraisal

Potential appraisal is a future oriented appraisal whose main objective is to identify and evaluate the potential of the employees to assume higher positions and responsibilities in the organizational hierarchy. Many organisations consider and use potential appraisal as a part of the performance appraisal processes

Indicators of Potential Appraisal

Indicators	*Interpretation*
1. A sense of reality	Extent to which a person thinks and acts objectively.
2. Imagination	The ability to let the mind range over a wide variety of possible causes of action, going beyond conventional approaches.
3. Power of analysis	The capacity to break down, reformulate or transform a complicated situation into manageable terms.

Indicators	*Interpretation*
4. Breadth of vision	The ability to examine a problem in the context of a much broader framework of reference to detect within a specific situation affecting the status.
5. Persuasiveness	The ability to sell ideas to other people and gain a commitment from individual to support 'management authority'.

Purpose of Potential Appraisal

The purposes of a potential appraisal are:

i. To inform employees of their future prospects
ii. To enable the organisation to draft a management succession programme
iii. To update training and recruitment activities
iv. To advise employees about enhancing career opportunities

Advantages of Potential Appraisal

i. The organizations are able to identify individuals who can take higher responsibilities.
ii. It also conveys the message that people are not working in dead-end jobs in the organization.
iii. The identification of employee potential to ensure the availability of people to do different jobs helps to motivate employees in addition to serving organisational needs.

Steps of Potential Appraisal System

I. **Role description**: A good potential appraisal system would be based on clarity of roles and functions associated with the different roles in an organisation. This requires extensive job descriptions to be made available for each job.

II. **Qualities required**: Besides job descriptions, it is necessary to have a detailed list of qualities required to perform each of these functions. These qualities may be broadly divided into four categories:
   i. Technical knowledge and skills
   ii. Managerial capabilities and qualities
   iii. Behavioural capabilities
   iv. Conceptual capabilities.

III. **Indicators of qualities**: A good potential appraisal system besides listing down the functions and qualities would also have various mechanisms for judging these qualities in a given individual. Some of the mechanisms for judging these qualities are:

i. Rating by others
ii. Psychological tests
iii. Simulation games and exercises
iv. Performance appraisal records.

IV. **Organising the system**: Once the functions, the qualities required performing these functions, indicators of these qualities, and mechanisms for generating these indicators are clear, the organisation is in a sound position to establish and operate the potential appraisal system. Such establishment requires clarity in organisational policies and systematisation of its efforts.

V. **Feedback**: If the organisation believes in the development of human resources it should attempt to generate a climate of openness. Such a climate is required for helping the employees to understand their strengths and weaknesses and to create opportunities for development. A good potential appraisal system should provide an opportunity for every employee to know the results of assessment.

A good potential appraisal system provides opportunities continuously for the employee to know his strengths and weaknesses. These are done through periodic counselling and guidance sessions by either the personnel department or the managers concerned. This should enable the employee to develop realistic self-perceptions and plan his own career and development.

C. Task Management System

Definition

A Task Management System is a structured approach to planning, organizing, tracking, and completing tasks or activities efficiently. It offers a centralized platform that helps people or teams manage their work efficiently. This platform is usually in the form of software. These systems let users create, allocate, prioritize, and monitor tasks to ensure nothing is missed.

Task management systems facilitate teamwork by enabling communication and information sharing among team members. They also include task classification, deadline setting, and progress updates. These systems are a vital resource for optimizing task-related activities as they help keep things organized, boost productivity, lower stress levels, and ensure responsibility in various personal and professional contexts.

The Importance of Task Management System

Effective task management is a key component of productivity and success in today's fast-paced world. Task management solutions are essential to achieve this efficiency and benefit both individuals and companies. These are, organization, periodization, time management, collaboration, decreased stress, accountability and flexibility

I. **Organization:** Task management systems offer a systematic structure that facilitates the organization of tasks. They enable users to make, group, and order tasks so that nothing is missed or forgotten.

II. **Prioritization:** The capacity of task management systems to prioritize work is one of their main advantages. Users can increase productivity by concentrating on the most important tasks and prioritizing each job.

III. **Time management:** Setting deadlines and allotting time for each activity are essential to effective task management. With the aid of this functionality, users may more efficiently manage their time, meet deadlines, and make the most of their priceless resources.

IV. **Collaboration:** Task management solutions facilitate collaboration and openness among team members in a professional setting. The system makes it simple to assign, monitor, and amend tasks, promoting cooperation and enhancing general project management.

V. **Decreased stress:** Anxiety and tension can arise from feeling overburdened by a never-ending list of things to do. Task management systems reduce stress and promote mental well-being by breaking down big tasks or many duties into smaller, manageable components.

VI. **Accountability:** Task management systems make it obvious who is in charge of what by explicitly assigning duties. Team members' adherence to task accountability is ensured via real-time tracking and status reports.

VII. **Flexibility:** Adaptable task management solutions are available. They allow users to modify their plans in response to evolving conditions, enabling them to handle unforeseen occurrences or priority changes.

Benefits of Task System

i. Increased productivity
ii. Enhanced cooperation
iii. Improved decision-making
iv. Increased accountability
v. Stress reduction
vi. Time savings

D. Development System

Training and Development in Human Resource Management is the process of acquiring knowledge, skills, and attitude that helps improve employees' job performance and enables future career growth. Training refers to acquiring specific knowledge and skills for a particular job or task.

Human Resource Development System

The combination of the four foci of HRD (Individuals, Dyads, Collectives, and Organisation) with four agents of HRD (Self, Immediate boss, HR department, and Organisations) gives the HRD systems. In matrix representing the HRD system is given herewith. As will be seen, the elements in the cells are the various development aspects. HRD system caters to all human resource units of the organizations, and all concerned are involved in running the system.

Faci of HRD	*Agents of HRD Process*			
	Person (self)	*Immediate boss*	*Professionals of HRD Dept.*	*Organi-zation*
The Individual				
i. Goal setting				
ii. Performance analysis				
iii. Performance				
iv. Improvement				
v. Humanisation,				
vi. Education				
vii. Advancement				
The Dyad (employee-boss)				
i. Trust				
ii. Mutuality				
iii. Helping				
The Collectives (teams)				
i. Effectiveness				
ii. Collaboration				
Department/Organisation				
i. Climate				
ii. Self-reviewing				
iii. Process				

E. Socialization System in HRM

Socialization clarifies new employees' roles, responsibilities, and performance expectations within the organization. It ensures that they understand their job duties, goals, and how their contributions contribute to the organization's success.

HRM in socialization refers to the process of integrating new employees into organization's culture, values norm and practices.

Purpose of Socialization

i. Cultural integration
ii. Role clarity
iii. Reducing turnover
iv. Enhancing performance
v. Building relationship
vi. Cultural reinforcement

Stages of Social Integration

Stages	*Description*
1. Pre-arrival stage	The stage indicates initial perception, expectation, preparation for new job role and environment. It may include integration with recruiters.
2. Orientation stage	After joining receive instruction about company's policy, culture, practices, procedure and administrative task.
3. Encounter stage	The new employees realise the situation, interaction with colloquies and try to be familiar with all sections.
4. Adjustment stage	New employees adapt to the rules, work environment, organization's culture, become more familiar with task.
5. Setting in stage	Become more comfortable and integrated to the organization.
6. Full integration stage	Full integration to the organization and feel as valued member of it.

Challenges

i. Cultural assimilation
ii. Information overload
iii. Time constraints
iv. Employees' turn over
v. Misalignment with expectations: May have unrealistic expectation leading to disappointment
vi. Technical and social aspects specific to the job and the company is essential to function in a new environment.

F. Governance in HRM

HR Governance is the bridge that seamlessly connects HR practices with organisational goals, ensuring that the workforce actively contributes to the realisation of business success. It goes beyond the traditional administrative role of HR by integrating human capital as a strategic asset.

Despite the seeming complexity of HR governance, management in the nonprofit sector can simplify the process by introducing four basic pillars as their governance framework: developing human resources policies, implementing and adhering to the policies, reviewing

Introduction

Navigating the intricate landscape of human resources within an organization requires a robust framework to ensure fairness, compliance, and strategic alignment. This is where HR Governance steps into the spotlight, serving as the compass that directs the course of an organization's human capital management.

What is HR Governance?

At its core, HR Governance refers to the set of policies, practices, and structures that guide and oversee the management of an organization's human resources. Unlike traditional HR management, HR Governance transcends day-to-day operations, focusing on strategic alignment, risk mitigation, and ensuring ethical practices throughout the employment lifecycle.

The Two Components of HR Governance

Within the intricate tapestry of HR Governance, a duality emerges, giving rise to two indispensable components that shape the very essence of how an organization manages its human resources.

1. *Strategic Component*

At the helm of HR Governance stands its strategic component, a visionary force that seeks to align human resource practices with the overarching goals and objectives of the organization. This is not a mere exercise in synchronization but a deliberate effort to position the workforce as a strategic asset, a catalyst for organizational success.

- **(i) Strategic workforce planning:** The strategic aspect involves forecasting future talent needs based on the organization's long-term goals. This requires an in-depth understanding of the business landscape, industry trends, and an anticipation of skill requirements. By aligning HR strategies with organizational objectives, companies can ensure that they have the right talent in the right place at the right time.
- **(ii) Talent management and development:** This component extends beyond recruitment and involves the holistic management and development of talent. It encompasses succession planning, leadership development, and initiatives that nurture a culture of continuous learning. Strategic talent management ensures that the workforce is equipped with the skills and capabilities needed to drive the organization forward.
- **(iii) Strategic HR metrics and analytics:** To measure the effectiveness of HR strategies, organizations leverage advanced metrics and analytics. This

involves tracking key performance indicators (KPIs) related to employee productivity, engagement, and retention. By collecting and analyzing data, HR professionals gain insights that inform strategic decisions and drive continuous improvement.

(iv) **A master class in strategic HR management:** One of the current buzzwords in business is HR Transformation. The idea is to ensure that we are driving business and getting results while improving the functionality of HR.

2. *Operational Component*

While the strategic component sets the direction, the operational facet of HR Governance is the machinery that ensures the seamless execution of policies and practices. It is the day-to-day engine that powers HR functions, maintaining compliance, fairness, and efficiency across the employment lifecycle.

(i) **Policy development and implementation:** At the core of the operational component is the development of clear and comprehensive HR policies. These policies cover a myriad of areas including recruitment, compensation, benefits, performance management, and employee relations. The key lies not only in creating these policies but in their consistent and fair implementation across all levels of the organisation.

(ii) **Compliance and legal adherence:** The operational side of HR Governance is intricately linked with legal compliance. HR professionals must stay abreast of labour laws, regulations, and industry standards to ensure that organizational practices adhere to legal requirements.

(iii **Efficient HR processes:** Streamlining HR processes is a cornerstone of the operational component. From on boarding to off boarding, payroll processing to performance appraisals, efficiency is key. Implementing technology solutions, standardizing workflows, and ensuring clear communication channels contribute to the operational excellence of HR Governance.

In summary, the operational component of HR Governance is the backbone of the framework, ensuring that the strategic vision is translated into actionable practices. It emphasizes the meticulous implementation of policies, compliance with legal standards, and the optimization of day-to-day HR operations. Together, these two components create a dynamic and cohesive HR Governance framework, positioning the organization for sustained success in the ever-evolving landscape of modern business.

What are the Four Pillars of HR Governance?

1. Policy development and implementation
2. Risk management

3. Compliance and ethics
4. Performance monitoring and improvement

To build a robust HR Governance framework, organizations must reinforce it with four essential pillars, each playing a pivotal role in sculpting the culture and efficacy of human resource management.

1. **Policy development and implementation:** The first pillar involves more than just the creation of policies; it's about crafting a living, breathing document that reflects the organization's values and commitments. This dynamic approach ensures that policies evolve alongside the business landscape, adapting to new challenges and opportunities while maintaining a foundation of fairness and consistency.
2. **Risk management:** Beyond risk mitigation, this pillar is the guardian of organizational resilience. It's about fostering a proactive approach to potential challenges, anticipating changes in the regulatory environment, and building a risk-aware culture where HR professionals become adept at navigating uncertainties, steering the organization clear of potential pitfalls.
3. **Compliance and ethics:** As the moral compass of HR Governance, this pillar extends beyond regulatory adherence. It encapsulates the organization's dedication to ethical behaviour, emphasizing the importance of integrity, transparency, and social responsibility. By weaving ethical considerations into the fabric of HR practices, organizations not only stay compliant but also earn the trust and respect of their workforce and stakeholders.
4. **Performance monitoring and improvement:** Beyond the metrics and evaluations, this pillar is the driving force behind organizational agility. It's about fostering a culture where continuous improvement is not a buzzword but a way of life. Statistics show that a whopping 98 per cent of organization.

Why is HR Governance Important?

Effective HR Governance is not just a bureaucratic necessity; it is the key to unlocking organizational success. Here's why it holds such paramount importance:

1. **Risk mitigation:** HR Governance serves as a shield against the multitude of risks associated with human resources. From legal compliance to financial prudence, the framework identifies, assesses, and mitigates potential risks, safeguarding the organization from legal entanglements, financial setbacks, and reputational damage. In essence, it fortifies the organizational structure, creating a resilient foundation that can weather uncertainties.
2. **Compliance and ethical practices**: Beyond the regulatory checkboxes, HR Governance is the custodian of organizational ethics. It ensures that

the organization operates within the bounds of legal requirements and ethical standards. This not only protects the organization from legal repercussions but also fosters a culture of integrity and transparency, crucial elements for building trust among employees, stakeholders, and the wider community.

3. **Operational efficiency:** The impact of HR Governance vibrates through the corridors of operational efficiency. By providing a structured framework for policy development, implementation, and compliance, it streamlines HR processes. This, in turn, reduces administrative burdens, minimizes the risk of errors, and enables the HR department to respond swiftly to the evolving needs of the workforce.

Steps to Foster Good Governance in HR

Implementing effective HR Governance requires a strategic approach. Here are the key steps organizations can take to foster good governance in HR:

1. Establish Clear Policies
2. Regular Training and Development
3. Create a Culture of Compliance
4. Implement Robust Risk Management Processes
5. Performance Monitoring and Continuous Improvement

In essence, fostering good governance in HR is a multifaceted endeavor that combines strategic vision with meticulous operational execution. It requires a commitment to transparency, adaptability, and the ongoing development of both policies and people. By taking these steps, organizations can cultivate a culture of excellence that not only meets the current needs of the workforce but also positions the HR function as a strategic driver of organizational success.

Metrics to Measure Good Governance in HR

Metric	*Description*	*Importance*
1. Employee Engagement	Staff satisfaction and organizational commitment	Indicates workplace well-being
2. Compliance Adherence	Tracks adherence to legal and ethical standards	Ensures risk mitigation and trust
3. Talent Retention Rate	Measures employee retention over a specific period	Reflects HR effectiveness in Retaining talent
4. Time-to-Fill Vacancies	Evaluates the speed of filling open positions	Effect efficiency in recruitment
5. Training Program Impact	Assesses the impact of training on employee performance	Indicates effectiveness of development initiatives

Key Resources to Help Implement Effective HR Governance

Building a strong HR Governance framework requires access to the right resources. Here are some key resources that organizations can leverage are:

i. Industry associations and networks
ii. Legal and compliance experts
iii. Technology solutions
iv. Training and development programs
v. Consulting services
vi. HR governance framework templates

In summary, effective HR Governance necessitates access to a diverse set of resources. Industry connections, legal expertise, technological tools, ongoing education, external consulting, and templates collectively form a toolkit that empowers organizations to navigate the multifaceted landscape of human resource management. By strategically leveraging these resources, organizations can fortify their HR Governance framework, ensuring it remains agile, compliant, and aligned with the ever-evolving demands of the workforce and the business environment.

G. Functions of HRD - Organizational Development

From Sources Across the Web

1. Performance management
2. Succession planning
3. Career development
4. Human Resource Development
5. Training and development
6. Development programs
7. Employee engagement and motivation
8. Leadership development
9. Organizational development
10. Career planning and development
11. Change management
12. Ensuring worker satisfaction
13. Increased productivity
14. Strategic talent acquisition
15. Talent management
16. Continuous improvement
17. Employee training
18. Ethical development

19. Evaluation
20. HR analytics
21. Improvement in worker performance
22. Personal development
23. Learning in the work environment
24. Organizational culture

H. Career Development

Career development refers to the processes organizations implement to support employees and encourage their career progression. Careers are rarely static. Skills, circumstances, desires, and motivations change over time and impact what people want out of their professional lives

Meaning of Career Development

Career development is the proactive, lifelong process of finding footing and advancing career path. It's an intentional approach to creating a meaningful career that includes setting long-term goals, exploring professional development opportunities, and gaining new work experience.

In today's job market, career development is increasingly important. With changes in technology and the global economy, professionals must adapt to stay competitive. A strategic **career development plan** helps grow future career opportunities.

How Career Development Works?

Career development can take different forms, but people generally take two paths: through an organization or independently.

Professionals who develop their careers through an organization tend to have more resources available. Their company may offer development programs, as well as a support network of employees.

Independent professionals often must look for career development opportunities on their own. They might use online resources like social media and online classes, attend conferences or networking events, meet with a career counselor, or join professional organizations and read trade publications.

Two types of career development are:

i. **Formal career development.** This includes short-term training programs, education, certifications, workshops, or seminars that can help build skill sets for a particular job or industry.

ii. **Informal career development.** This includes mentorship opportunities, networking events, online courses, internships, and volunteering experiences.

Importance of Career Development

Having a structured career development plan helps professionals ensure that they are entering and traversing a career path that makes the most of both their skills and values.

1. Helps make use of skills
2. Increases motivation at work
3. Helps achieve financial goals
4. Increases employee retention
5. Increases happiness and satisfaction
6. Identify and encourage unique skills
7. Promote cross-functional projects
8. Provide networking opportunities
9 Offers technology and tools.

I. Capacity Development

Capacity-building is defined as the process of developing and strengthening the skills, instincts, abilities, processes and resources that organizations and communities need to survive, adapt, and thrive in a fast-changing world.

What are the 6 Pillars of Capacity Development?

1. Structure
2. Competencies
3. Management systems
4. Enabling policies
5. Knowledge and learning and lastly
6. Leadership

The IMF provides technical assistance and training known as capacity development as one of its core functions. Capacity development accounts for around a third of the IMF's annual spending. It is available to all members upon their request and is tailored to a country's specific needs. Capacity development can help countries to improve tax collection and bolster public finances. It can help countries to modernize their monetary and exchange rate policies, develop legal systems, or strengthen governance. Capacity development also includes capacity building approaches and best practices.

Capacity Development Strategies

1. *Up skilling employees*

(i) **Up skilling** is the process of improving an employee's work performance by building on their current skills. For example, a marketer might learn how to utilize digital tools to evaluate campaign performance.

(ii) **Reskilling** involves training employees to transition to new roles by teaching them skills unrelated to their current position. An example would be a marketer changing their role into a data analyst.

2. *Leadership Development*

Effective leaders are at the heart of creating a culture of continuous improvement and professional growth. They can help:

i. Identify the unique strengths and needs of their team members
ii. Monitor their growth and motivate them
iii. Maintain organization continuity with succession planning

Ultimately, investing in leadership capacity development ensures that an organization has the guidance and strategic oversight needed to address challenges and capitalize on growth opportunities.

3. *Outsourcing Resources*

Outsourcing resources can be a good strategy to enhance capacity at short notice. It allows organizations to quickly access specialized skills and expertise that may not be available internally, filling gaps without the time and expense of recruiting and training new employees

4. *Building Capacity through Partnerships*

By seeking out like-minded agencies to collaborate with, one can fill gaps in service delivery to deliver additional value to clients. Partnerships help agencies combine their resources, share knowledge, and create something new that benefits both sides.

Consider some best practices that Nick Dan-Bergman, the VP of Marketing and Partnerships at Lane Terralever, shared with Productive Bold Community:

i. To start off, consider establishing partnerships in area of expertise
ii. Find influential companies that have some skin in the game
iii. Figure out who's the right point of contact for company
iv. Manage expectations this goes for internal teams as well
v. Be patient

5. *Development Plan for Agency Processes*

Streamlining communication is an important part of efficient service delivery and balanced workloads. According to research by McKinsey, employees spend up to 10 hours per week searching for the information they need to do their tasks. Adopting collaborative tools with real-time sharing and data updating can also ensure that team members have immediate access to the resources they need.

Definition and Meaning of the Terms Used

A

1. **Abstract:** In management, an abstract is a brief summary of a presentation topic, paper, or poster that is submitted for consideration at a conference or event. The process of managing abstracts for a conference is called abstract management.
2. **Academic Strategy:** Academic strategy is the systematic planning and implementation of methodologies to improve academic performance and institutional quality. It involves using tools like the Balanced Scorecard (BSC) method and Multiple Criteria Decision Analysis (MCDA) techniques.
3. **Accessibility:** Accessibility in management refers to the process of creating environments that are inclusive and allow everyone to participate in an organization's programs, services, and activities. This includes physical, digital, and communication accessibility.
4. **Accountability:** Accountability in management is the expectation that employees and managers are responsible for their actions, decisions, and performance. It's a key component of organizational success, and can lead to increased productivity, employee motivation, and commitment to work.
5. **Action Research:** Action research is a research methodology that involves taking action and conducting research simultaneously to create change. It's often used in the social sciences to address real-world problems in a collaborative and cyclical way.

6. **Activity Plan:** An activity plan in management is a process of organizing and thinking about the activities required to achieve a goal. It involves defining the activities that need to be completed to bring a project to fruition.
7. **Activity Strategy:** An activity strategy is a set of actions that are taken to achieve a broad concept or approach, known as a strategy. For example, a strategy to build the capacity of community members might include a variety of activities.
8. **Agility:** Workplace agility is about employees working smarter and more collaboratively, productively and effortlessly not harder. In short, agile employees (and therefore their companies) are willing and able to move and adjust quickly and easily to change, as often as necessary.
9. **Analytical Reporting:** Analytical reporting is a process that involves creating a written document that analyses a specific issue, data set, or scenario to help businesses make decisions, solve problems, and identify opportunities. Analytical reports are important because they help businesses make informed decisions based on evidence, rather than intuition.

B

1. **Barometric Method:** This method is based on the past demands of the product and tries to project the past into the future. The economic indicators are used to predict the future trends of the business. Based on future trends, the demand for the product is forecasted. An index of economic indicators is formed.
2. **Baseline Data:** Baseline data often describes a situation that existed before an event, which can be defined in many ways, depending on the operational context. An event might be a drought or an incident of political upheaval, or it may simply be the first time the indicators were ever measured.
3. **Bias Reduction:** Bias is having a preference for something over another thing. The Law of Attraction is research that supports the idea that everyone has biases, even if they are often implicit. Ways to reduce bias towards something are to identify biases, pursue empathy, increase diversity, and consciously act.
4. **Buzz Session:** It is a complete evidenced-based social and emotional skills program used in more than 130 countries to teach young people social and emotional intelligence. This amazing program teaches a range of essential interpersonal skills, simulated by the experiences of a character called Archie.

C

1. **Capacity Building:** Capacity building is the process of developing and strengthening the skills, abilities, and resources of organizations and

communities to adapt, survive, and thrive in a changing world. It can help teams achieve their objectives, manage challenges, and contribute to organizational growth.

2. **Card Sort Method:** The card sort method is a technique used to organize information and understand how people think about it. It's commonly used in management to assess and design the structure and navigation of websites and applications.
3. **Career Development:** Career development is the process of helping employees grow professionally and progress in their careers.
4. **Case Study:** A case study is a research approach that is used to generate an in-depth, multi-faceted understanding of a complex issue in its real-life context.
5. **CD Intervention:** Capacity development interventions in management are a process to help individuals, organizations, and systems improve their abilities to achieve goals and contribute to positive change. Capacity development can be applied to a variety of sectors and goals.
6. **Centralized Training:** Centralized training is a training model where a company's training and development is managed by a single group or department. This model is often used in smaller organizations where resources are not spread out across multiple lines of business or locations.
7. **Competence:** Competencies are the knowledge, skills, abilities, and behaviours that contribute to individual and organizational performance.
8. **Complaint:** Complaint management is the process of handling customer complaints to improve customer satisfaction and loyalty, and to protect a company's bottom line.
9. **Conference:** A conference is a meeting where people gather to discuss a specific topic or share information. Conferences can be held for a single day or over several days, depending on the event's size and scope.
10. **Consultant:** Management consulting is a global industry focused on helping organizations operate effectively. A management consultant, or management analyst, provides an outside perspective on problem-solving, best practices, and strategy to help companies improve their performance.
11. **Content Input Reaction Output (CIRO):** Content Input Reaction Output (CIRO) is a model for evaluating the effectiveness of management training programs. The model is hierarchical, with four stages.
12. **Counselling:** Counselling in management is a proactive people management strategy that can help employees manage stress, personal issues, and work-related problems. It can also help organizations reduce costs associated with turnover, absenteeism, and accidents.

13. **Crafting Model:** Job crafting is a model in human resources development (HRD) that allows employees to define their roles and responsibilities. It's a way to make a current job more engaging and satisfying for the employee.
14. **Credibility:** Credibility in human resources (HR) is the quality of being trustworthy and believable.
15. **Critical Incident Technique:** The critical incident technique is frequently used to collect data on observations previously made which are reported from memory. This is usually satisfactory when the incidents reported are fairly recent and the observers were motivated to make detailed observations and evaluations at the time the incident occurred.
16. **Critical Reflection Action Research:** Critical reflection is the process by which adults identify the assumptions governing their actions, locate the historical and cultural origins of the assumptions, question the meaning of assumptions, and develop alternative ways of acting (Cranton, 1996).
17. **Cross Functional Training:** Cross-functional training is a talent development practice that helps employees learn new skills and knowledge to work in different areas of an organization.
18. **Cross Training:** Cross-training is a professional development technique that involves teaching employees to perform tasks or roles outside of their primary job duties.
19. **Culture:** Culture in Human Resource Management (HRM) refers to the set of shared values, beliefs, norms, behaviors, and customs that shape the way people in an organization interact, make decisions, and work together.
20. **Culture System:** In Human Resource Development (HRD), a culture system is a climate that establishes values, norms, and culture, and ensures a high level of motivation. The goal of a culture system is to create an acceptable organizational culture that is conducive to work.
21. **Customization:** Custom development involves creating software from scratch or modifying existing systems to meet unique requirements, offering tailored functionality, scalability, and integration capabilities. It provides a competitive edge but can be costlier and time-consuming.

D

1. **Data Store:** A data store is a digital repository that stores and safeguards the information in computer systems. A data store can be network-connected storage, distributed cloud storage, a physical hard drive, or virtual storage.
2. **Decision Making:** Decision making is the process of choosing a course of action or belief from a set of possible options. It can be a cognitive process that involves rational or irrational thought.

3. **Delphi Method:** The Delphi method is a process that uses a panel of experts to reach a group consensus on a complex issue. It involves a series of rounds of questionnaires, where experts anonymously provide their opinions and then adjust their answers based on the group's responses. The goal is to establish a reliable consensus that represents the group's collective intelligence. The Delphi method was developed by RAND in the 1950s to forecast the effect of technology on warfare. It's typically administered by email or virtually.
4. **Design:** Organizational design is the process of structuring an organization to achieve its goals and objectives. It involves aligning people, their skills, and the work they do to improve efficiency and effectiveness.
5. **Development:** Development is a process that creates growth, brings in progress and positive change.
6. **Diagnostic Assessment:** Diagnostic assessments in management can be used to evaluate an organization's capabilities, identify strengths and weaknesses, and plan for improvement.
7. **Diagonal Communication:** Diagonal communication, also known as crosswise communication, is a type of business communication that allows people from different departments and hierarchical levels to communicate directly. It's a combination of vertical and horizontal communication, which involves employees within the same rank or a supervisor and employee within the same team.
8. **Downward Communication:** Downward communication occurs when information and messages flow down through an organization's formal chain of command or hierarchical structure. In other words, messages and orders start at the upper levels of the organizational hierarchy and move down toward the bottom levels.

E

1. **Empathy:** Empathy is the ability to understand and share the feelings of another person.
2. **Empirical Method:** The empirical method is a research procedure that uses experimentation and observation to gain knowledge and test or refine theories.
3. **Employee Engagement:** Employee engagement is the level of commitment and enthusiasm that an employee feels towards their organization, work, and colleagues. It's a psychological state that's different from job satisfaction, employee behavior, or management action.
4. **Employee Satisfaction:** Employee satisfaction refers to how happy and fulfilled employees are with their jobs, encompassing their overall experience and contentment with their roles and the organization. It is a key metric for gauging the overall health of an organization.

5. **Employee Segment:** Employee segmentation, also known as workforce segmentation, is a method of dividing employees into groups based on shared characteristics. These groups are called segments. The goal of employee segmentation is to help employers understand their employees better and focus resources on what will have the most impact.
6. **Enabling Environment:** An enabling environment is a space that is supportive, safe, and rich with opportunities for learning and growth.
7. **Ethical Behaviour:** Ethical behaviour is characterized by honesty, fairness and equity in interpersonal, professional and academic relationships and in research and scholarly activities. Ethical behaviour respects the dignity, diversity and rights of individuals and groups of people.
8. **Extrinsic Reward:** Extrinsic rewards are incentives that are given to motivate people for completing a task or achieving a goal. They are external to the work itself and are controlled by people other than the employee.

F

1. **Feasibility Analysis:** A feasibility analysis is an objective assessment of the viability of a project or idea. It helps determine if a project is possible and how to make an informed decision about it.
2. **Financial Analysis:** Financial analysis is the process of examining a company's performance in the context of its industry and economic environment in order to arrive at a decision or recommendation.
3. **Flash Mentoring:** Flash mentoring is a short-term or one-time mentoring session where participants learn a skill or piece of information from an expert. It's a flexible way to get guidance on a specific challenge or skill, and can be used in a variety of settings, including the workplace and in professional associations.
4. **Focus Group Discussion:** A focus group discussion (FGD) is a qualitative research method that involves a group of people discussing a topic of interest in an interactive setting.
5. **Follow up:** Follow up means to take further action, gather more information, or maintain contact with someone to monitor the effects of previous activities.
6. **Formal Induction:** Formal induction training, also known as orientation, is a formal process that introduces new employees to a company and their role. It typically includes: orientation sessions, presentations, training modules, and equipment training.
7. **Formative Assessment:** Formative assessment is a way for teachers to evaluate how students are learning and improve their teaching and learning activities. It's a collaborative process that involves teachers and students, and

is designed to help students understand their learning goals and become self-directed learners.

8. **Functional Capacity:** Functional capacity refers to the capability of performing tasks and activities that people find necessary or desirable in their lives.
9. **Functional Intelligence:** Functional intelligence is the ability to understand and apply complex systems and concepts. It's the power to think critically, solve problems creatively, and make sound decisions under pressure.

G

1. **Gap Analysis:** A gap analysis is a process that compares a company's current performance to its desired performance. It can help a company identify areas where it needs to improve, and create a plan to achieve its goals.
2. **Generic Brand:** A generic brand is a product that is sold without a brand name or logo, and is usually identified by its packaging or product characteristics. Generic brands are often less expensive than brand name products and are designed to be a substitute for them.
3. Group Mentoring: Group mentoring is a mentoring strategy where a single mentor works with multiple mentees at the same time. In this format, mentees with similar goals or needs can learn and grow together, and the mentor can share their expertise and knowledge.

H

1. **Honesty:** Honesty is a moral character trait that involves being truthful, trustworthy, loyal, fair, and sincere. It also means being free of lying, cheating, and theft.
2. **Horizontal Communication:** Horizontal communication is the exchange of information between people or departments at the same level in an organization. It's also known as lateral communication.
3. **Horizontal Promotion:** A horizontal promotion is when an employee moves to a new role at the same level within an organization, while keeping the same overall rank and authority. This type of promotion can involve a change in job title and responsibilities, but the employee's level remains similar.
4. **Horizontal Rotation:** Horizontal job rotation is a process where employees move between different roles or positions at the same hierarchical level within a company. It's also known as a transfer.
5. **Hostile:** Some synonyms are inimical, antagonistic, unfavorable, and unfriendly. Definitions of hostile. Adjective. Characterized by enmity or ill will "a hostile nation".

6. **HR Governance:** HR governance is the process of managing human resources in a way that is consistent with legal, ethical, and social standards. It's a part of corporate governance that involves the internal and external framework of human resource management.

I

1. **In-Service Training:** In-service education, also called in-service training (IST), in-service training, education, and practice (ISTEP), teacher professional development (TPD), and continuing education, refers to the professional development of teachers who are already in the classroom.
2. **Induction Training:** Induction training (also known as on boarding/ orientation) is the process of introducing a new employee to their new role within the business. The purpose of this training is to help new starters become familiar with not only their role but what is expected of them as a member of your workforce.
3. **Informal Induction:** Informal induction relies on a more casual, ad-hoc approach. New employees may learn about their roles and the organization through interactions with colleagues and supervisors without a structured program.
4. **Informative Approach:** Informative approach is merely, formal, non-interactive that pushes the information to the learners; therefore can also be named as push-approach. Many e-Learning courses rely on heavy content with little or no interaction to make the course appealing, quicker, informative, and cost-effective.
5. **Input-output Analysis:** Input-output analysis (I-O) is a macroeconomic analysis based on the interdependencies between different economic sectors or industries. Input-output analysis is used to estimate the impacts of positive or negative economic shocks and analyzes the ripple effects throughout the economy.
6. **Internal Sourcing Data:** Some of the major internal data sources in a business include sales, marketing, finance, and human resources. Sales deal with information on products, marketing in web traffic, finance in the revenues and profits, and human resource in employee relations.
7. **Internship Training:** An internship is a professional learning experience that offers meaningful, practical work related to a student's field of study or career interest. An internship gives a student the opportunity for career exploration and development, and to learn new skills.
8. **Intrinsic Reward:** Intrinsic rewards are those which are internally driven by enjoyment for the meaningful work tasks being carried out. Intrinsically motivated employees work at higher levels of productivity and strive to improve their professional development skills.

J

1. **Job Analysis:** The meaning of job analysis is the practice of gathering and analyzing details about a particular job, such as responsibilities, day-to-day duties, hard and soft skills, qualifications, education, expected outcomes, interaction, performance standards, work conditions, physical abilities and supervision.
2. **Job Enrichment:** Job enrichment involves creating meaningful roles through feedback, encouragement, communication, and autonomy.
3. **Job Rotation:** Job rotation is the practice of moving employees between jobs in an organization. These rotations are predominantly lateral, meaning that they happen between jobs on the same level and are not considered promotions.
4. **Job Satisfaction:** Job satisfaction, employee satisfaction or work satisfaction is a measure of workers' contentment with their job, whether they like the job or individual aspects or facets of jobs, such as nature of work or supervision.
5. **Job Enrichment Model:** Job enrichment is a strategy used to motivate employees by giving them increased responsibility and variety in their jobs. The idea is to allow employees to have more control over their work

K

1. **Kirkpatrik Model:** This model is globally recognized as one of the most effective evaluations of training. The Kirkpatrick model consists of 4 levels: Reaction, learning, behaviour, and results. It can be used to evaluate either formal or informal learning and can be used with any style of training.
2. **Knowledge Management:** Knowledge management is the process of creating, curetting, sharing, using, and managing knowledge across an organization and even across industries.

L

1. **Lateral Rotation:** Lateral rotation, also known as external rotation, is a rotational movement away from the body's midline. It's the opposite of medial (internal) rotation, which is when the anterior surface of a limb moves toward the midline.
2. **Leadership:** Leadership is the ability to guide, influence, or lead a group of people or an organization. It's a set of behaviours and mind-sets that help people work together to achieve shared goals and adjust to change.
3. **Learning Retention:** Learning retention is the ability to store new information in long-term memory so that it can be recalled and used in the future.

4. **Lecture Method:** The lecture method is a traditional and widely used teaching strategy in which an instructor delivers a structured presentation of information to a large number of students. This approach is cantered around the instructor's verbal delivery.
5. **Life Mentor:** A Life Mentor on the other hand will have used the same questioning techniques used in life coaching to establish the client's goals, options and current position but will in addition be prepared to offer specific advice based on his or her own experience.

M

1. **Man Power Planning:** The process of deciding how many people will be needed for a particular job or project, and how they should be used: Man power planning consists of putting the right number and right kind of people in the right place.
2. **Management by Objective (MBO):** MBO is an acronym for Management by Objectives. It can be defined as a management system that measures employees' performance against a series of set targets or goals to gauge their overall performance in their role.
3. **Matched Comparison:** When using Matched Comparisons, participants (individuals, organizations or communities) are each matched with a non-participant on variables that are thought to be relevant which can be difficult to adequately match on all relevant criteria.
4. **Matching Process:** The matching process assigns numeric scores called weights to the comparison of individual data elements. The scores measure the contribution of each data element to the overall or composite weight. The composite weight, in turn, is the sum total weight of all defined comparisons.
5. **Maximized Customer Value:** Providing customers with the products or services they seek, at competitive prices and with terrific customer service is obviously the best way to ensure that a customer remains with a company for the long term and continues to buy from it.
6. **Mentor System:** Mentorship is a professional, working alliance in which individuals work together over time to support the personal and professional growth, development, and success of the relational partners through the provision of career and psychosocial support.
7. **Merit Based Promotion:** An organization seeking to promote employees may do so based on seniority to take advantage of the employees' experience with the organization and to reward them for their loyal service. Another option is merit-based promotion, which measures and rewards employees' contributions based on performance.
8. **Mission:** Any important task or duty that is assigned, allotted, or self-imposed.

9. **Mosaic Mentoring:** Mosaic mentoring is a new approach that emphasizes utilizing a collection of mentorship approaches to maximize outcomes based on individual and/or domain-specific needs.

N

1. **Need Analysis:** A training needs analysis (TNA) is a process that helps organizations determine the type of training that employees need to improve their job performance. It's a critical step in developing a training program, and it helps ensure that time, money, and effort are not wasted on unnecessary training.
2. **Non-Formal Training:** Non-formal training is a structured learning experience that takes place outside of formal learning environments, but within an organizational framework.

O

1. **On-boarding Training:** On boarding is the process by which new employees acquire the skills, knowledge and behaviours to become effective contributors to an organization. On- boarding programs are designed to acclimate individuals who are newly hired or transferred from within to new positions.
2. **Opinion Pool:** An opinion poll, often simply referred to as a survey or a poll is a human research survey of public opinion from a particular sample.
3. **Optimized Customer Budget:** An optimized customer budget is a marketing budget that is allocated to maximize the impact of spending and navigate diminishing returns.
4. **Organizational Capacity Building:** Organizational capacity building is a process that helps organizations improve their skills, knowledge, and resources to achieve their goals. It's a vital process for organizations to stay competitive and adapt to a changing world.
5. **Organizational Development:** Organizational development (OD) is a scientific approach that helps organizations improve their performance and culture through the implementation of new systems, practices, and techniques.

P

1. **Participation Counselling:** Participation counselling, also known as participative counselling, is a collaborative approach to counselling that involves a cooperative exchange of ideas between a counsellor and a counselee. It is neither completely counselor-centered nor completely counselee-cantered.

2. **Participatory Approach:** A participatory approach is a way of working that involves people who are directly affected by the outcome of a project or initiative.
3. **Peer Appraisal:** Peer appraisal is a performance evaluation method where employees are assessed by their colleagues. It's also known as peer assessment or peer evaluation.
4. **Peer Mentoring:** Peer mentoring is a type of mentorship where a person with experience helps a person who is new to that experience.
5. **Performance Based Approach:** Performance-based management is a systematic approach to performance improvement through an ongoing process of establishing strategic performance objectives; measuring performance; collecting, analyzing, reviewing, and reporting performance data; and using that data to drive improvement.
6. **Personal Development Approach:** Personal development is a lifelong process of improving self-awareness, knowledge, and skills.
7. **Persuasiveness:** Persuasiveness is the ability to convince others to do or believe something through argument or entreaty.
8. **Philip's ROI Model (Return on Investment):** The Phillips ROI Model is a training evaluation model that uses cost-benefit analysis to calculate the return on investment (ROI) of a program. The model is an extension of the Kirkpatrick model and adds a fifth level of evaluation.
9. **Potential Appraisal:** A potential appraisal is a process that evaluates an employee's potential to take on more responsibilities or higher positions in the future. It's also known as a future-oriented appraisal.
10. **Pre-employment Training:** Pre-job training, also known as pre-employment training, helps job seekers prepare for their new roles by teaching them new skills and improving existing ones. It can also help them understand workplace standards and company culture.
11. **Pre-invest Appraisal:** Pre-investment financial analysis allows to determine the investment potential, safety, timing and cost of return on investment of a project or a company, as well as to estimate further profit in the long term perspective.
12. **Problem Analysis:** Problem analysis is a process that helps identify the key issues of a problem, and how to best apply resources to solve it. It's a crucial skill for creative strategists, and can be used in a variety of contexts, including HR and software development.
13. **Promotion:** "What is Promotion?" Promotion is a marketing tool, used as a strategy to communicate between the sellers and buyers.

14. **Psychometric Test:** Psychometrics is a scientific discipline concerned with the question of how psychological constructs (*e.g.*, intelligence, neuroticism or depression) can be optimally related to observables (*e.g.*, outcomes of psychological tests, genetic profiles, neuro-scientific information).

Q

1. **Qualitative Method:** Qualitative research is a method of inquiry that uses non-numerical data to understand people's social reality.
2. **Quality Circle:** A quality circle is a group of employees who regularly meet to identify, analyse, and solve work-related problems. Quality circles are also known as quality control circles (QC) or Kaizen circles.
3. **Quality of Working Life:** Quality of work-life or QWL can be defined as the total quality of an employee's work-life at an organization. Not only QWE is tied to happier employees but also better business results. When the quality of work-life is stable, productivity is bound to increase and level of employee retention.
4. **Quantitative Method:** A quantitative method is a research approach that uses statistical analysis and numerical measurements to understand phenomena. It involves the following steps: Designing experiments or surveys, collecting data that can be quantified. Analyzing the data using mathematical models, and Drawing conclusions or making predictions.

R

1. **Rapport Building:** Rapport building is the process of developing a relationship or bond with someone to create trust and connection.
2. **Rational Method:** The rational decision-making model is a systematic method for making decisions that is often used by managers and leaders. It involves using logic and facts to reduce bias and emotion, and to make informed decisions.
3. **Recognition:** The identification of something as having been previously seen, heard, known, *etc.*, the perception of something as existing or true; realization, the acknowledgment of something as valid or as entitled to consideration.
4. **Recruitment:** Recruitment refers to the process of identifying, attracting, interviewing, selecting, hiring and on-boarding employees.
5. **Refresher Training:** Refresher training is a type of training that helps employees update their knowledge and skills in a subject they've already learned. It's also known as retraining or recurrent training.
6. **Reliability:** Reliability in research is a concept describing how reproducible or replicable a study is. In general, if a study can be repeated and the same results

are found, the study is considered reliable. Studies can be reliable across time and reliable across samples.

7. **Rescaling:** In management, rescaling is the process of changing the size or scope of a business in response to market or industry changes. It can involve: Downsizing, Restructuring, Shifting the business's focus, Scaling back operations, and exiting certain markets.
8. **Resilience:** In management, resilience is the ability of an organization to adapt to and recover from challenges, disruptions, or crises, while maintaining its core values and purpose.
9. **Retention:** Training retention refers to the ability to absorb and recall information learned during training. It's important for both the individual and the organization, as it can improve performance and bottom line.
10. **Retraining Training Cycle:** A training cycle is a program designed over a set period of time. The cycle organizes specific training elements within this time frame. For example, you may run increased distances or lift heavier weeks over a 6-week period.
11. **Reverse Mentoring:** Reverse mentoring is a mentoring practice where a junior employee mentors a more experienced colleague.
12. **Reward Management:** It is a systematic approach to motivating employees to achieve organizational goals. It involves creating and implementing policies and strategies to fairly and consistently reward employees based on their value to the organization.
13. **Risk Management:** Risk management is the process of identifying, assessing and controlling threats to an organization's capital, earnings and operations.
14. **Risk Mitigation:** Risk mitigation is the practice of reducing the impact of potential risks by developing a plan to manage, eliminate, or limit setbacks as much as possible.
15. **Role Analysis:** The main purpose of the analysis is to understand and interpret data to be used to make informed decisions.
16. **Role Efficiency:** Role efficacy is the potential effectiveness of an individual in a specific role within an organization. It's based on how well a person understands their role and responsibilities, and how they integrate their roles into the organization.
17. **Role Play:** Role-play is an active learning technique that helps people learn and develop skills by simulating real-world situations.

S

1. **Self-renewal System:** A self-renewal system in management is a process that helps an organization learn and develop, and to initiate and implement change programs without outside help.

2. **Self-training Mechanism:** Self-training is a simple and effective semi-supervised learning classification method.

3. **Semi-structured Training:** Semi-structured training programs are a type of training program. Training is a way to improve employee skills, knowledge, and development opportunities.

4. **Seminar:** A seminar is a meeting for discussing and sharing information, or a small course that focuses on discussion. Seminars can be academic, professional, or other types of events.

5. **Shadowing:** The shadowing technique is a method for learning to speak more fluently by mimicking the speech of a model.

6. **Skill Enhancement:** By continuously improving their competencies, individuals can enhance their skills and become more effective in their professional roles. Skill enhancement involves various aspects such as technical skills, soft skills, and transferable skills, all of which contribute to an individual's overall competency.

7. **Smoothing Technique:** It is a short-term forecasting technique that is frequently used in the production and inventory environment, where only the next period's value is required to be forecast. Because only three numbers are required to perform exponential smoothing, this technique is simple to update.

8. **Social Cost-Benefit Analysis (SCBA):** Social Cost-Benefit Analysis (SCBA) is a method that compares the total social benefits of a project to its total social costs to determine if the project is justified. It's a systematic tool that's used to evaluate long-term decisions for capital projects, and is based on welfare economic theory.

9. **Space Repetition:** Spaced repetition is a learning method that involves reviewing material at increasing intervals of time.

10. **Stagnation:** HR professionals, like those in any field, can experience periods where career growth seems to plateau.

11. **Strategic Alignment:** Strategic alignment is a management framework that ensures an organization's goals, resources, and activities are coordinated to work toward the same business goals.

12. **Strategic Component:** The main components of strategy management are goal setting, environmental analysis, strategy development, implementation and control. Goal setting involves setting clear, measurable objectives for the organization.

13. **Strategic Mind Set:** A strategic mind set is stepping out of the everyday details and 'doing', to look at the situation from above, giving an objective perspective through a bird's eye view. There are three parts to developing a strategic way of working and thinking.

14. **Strategic Planning:** Strategic planning is a process in which an organization's leaders define their vision for the future and identify their organization's goals and objectives. Structure Report: Generally, a report will include some of the following sections: Title Page, Terms of Reference, Summary, Table of Contents, Introduction, Methods, Results, Main body, Conclusion, Recommendations, Appendices, and Bibliography.
15. **Succession Development:** Succession development is a method used to identify and develop emerging leaders so that they can take control in time of need and keep a company afloat and unaffected.
16. **Succession Planning:** Succession planning is the process of identifying the critical positions within organization and developing action plans for individuals to assume those positions.
17. **Symposium:** A meeting or conference for the discussion of some subject, especially a meeting at which several speakers talk on or discuss a topic before an audience.
18. **SWOT:** SWOT stands for strengths, weaknesses, opportunities, and threats, and is a planning tool used to assess an organization's or project's internal and external factors. SWOT analysis is a strategic planning technique that can help identify areas for improvement, uncover blind spots, and develop a successful strategy.
19. **Syndicate:** The syndicate method is a training method that involves splitting a group of participants into smaller groups, called syndicates, to work on a project, exercise, or discussion. The method is used to expose participants to group dynamics, and to help them develop leadership skills and self-learning capabilities.
20. **Synergy:** A synergy plan is a roadmap that outlines how the combined entity will achieve greater performance, efficiency, and innovation than the sum of its parts.

T

1. **Talent Development:** Talent development is the process by which organizations ensure their workforce possesses the right skill sets and abilities to execute their jobs today and be prepared for changes to those jobs in the future.
2. **Task Force:** In government or business a task force is a temporary organization created to solve a particular problem.
3. **Team Mentoring:** Group mentoring, also known as team mentoring is where one individual acts as a mentor and oversees multiple mentees, it also can be reversed where a singular mentee has various mentors. Essentially, it's a

compilation of mentoring relationships coming together regularly for a set time.

4. **Techno-Economic Analysis:** Techno-economic analysis, or TEA, is a method for evaluating the economic performance of a technology. A TEA assesses the overall value of a technology, allowing analysts to objectively weigh benefits against costs.
5. **Techno-Structural Intervention:** Techno-structural interventions are targeted toward structural and technological issues such as organizational design, work redesign, and employee engagement.
6. **Theory:** A theory is a well-substantiated explanation of an aspect of the natural world that can incorporate laws, hypotheses and facts.
7. **Third Party Peace Making:** The third party will intervene directly and indirectly in facilitating the dialogue between the principles of the underlying factors at work and the feelings of the persons involved. Consequently, the resolution of the conflict requires open confrontation with another in the presence of the consultant.
8. **Time Management:** Time management is the process of organizing and planning how to divide your time between different activities.
9. **Time Series Analysis:** Time series analysis is a specific way of analyzing a sequence of data points collected over an interval of time. In time series analysis, analysts record data points at consistent intervals over a set period of time rather than just recording the data points intermittently or randomly.
10. **Traditional Research:** Traditional Research refers to the conventional research approach where investigator-driven projects are sponsored and led by researchers at academic institutions, aiming to provide evidence to inform decision-making in various fields.
11. **Training:** It is teaching, or developing in oneself or others, any skills and knowledge or fitness that relate to specific useful competencies.
12. **Training Goal:** SMART goals are Specific, Measurable, Achievable, Relevant, and Time-Bound. Each element of the SMART framework plays a role in enhancing the effectiveness of professional development initiatives.
13. **Training Need Assessment:** The purpose of a training needs assessment is to identify performance requirements and the knowledge, skills, and abilities needed by an agency's workforce to achieve the requirements.
14. **Training Objective:** Training or learning objectives are the intended measurable outcome that learners will achieve once they've finished a course.
15. **Training Participant:** Training Course Participant means a person who personally takes part in the Training Course. The Training Course Participant may be a person different from the Customer.

16. **Trend Analysis:** Trend analysis is a systematic process that uses statistical techniques to identify historical patterns and project future outcomes based on that data.

U

1. **Unstructured Training:** Unstructured learning refers to a learning experience that is not planned and does not have an organized progression.
2. **Upward Communication:** Upward communication is the process by which company employees communicate and share their thoughts and feedback with higher level management.

V

1. **Validity:** Validity refers to how accurately a method measures what it is intended to measure. If research has high validity, which means it produces results that correspond to real properties, characteristics, and variations in the physical or social world.
2. **Vertical Mentoring:** Vertical mentoring most of us know; it involves someone who is more experienced helping to guide one or more individuals who wish to learn from that experience.
3. **Vertical Promotion:** This refers to an upward movement of employees with a change in skills and experience. It brings a change in salary, responsibility, status, benefits, *etc.* In the marketing industry, this can be the promotion of a marketing supervisor to the marketing manager.
4. **Vertical Rotation:** Employees are rotated- to positions at different hierarchical levels within the organization. It allows them to gain a broader understanding of different levels of responsibility and develop leadership capabilities.

W

1. **Wellness Intervention:** By definition, a wellness intervention is a combination of program elements or strategies designed to produce behaviour changes or improve health status among individuals or an entire population.
2. **Work Environment:** A work environment is a space physical and emotional in which employees perform their daily tasks. It's a combination of three vital components: Physical environment: This covers the layout and amenities of an in-person office, including items like desk space, lighting and location.
3. **Work Load:** The amount of work that a person or machine has to do within a particular period of time. Unmanageable workloads can be a source of stress. They face increased workloads.

References

1. CIDA (1997), Guide to Gender Sensitive Indicators, Canada.
2. Kirkpatrick, L. D. and Kirkpatrick, D. J. (2005), Evaluating Training Programs–The Four Levels, Benett Koehler Publishers Inc.
3. Laird, D. (1978), Approaches to Training and Development/Reading, Mass, Addison-Wesley Publishing Company.
4. Hanna, G. S., and Dettmer, P. A. (2004). *Assessment for effective teaching: Using context-adaptive planning.* Boston, MA: Pearson A and B.
5. Marcotte, P. L., Bakker D. R. and Bell, M. (2002), Assessing Training Impact: IRRI's New Evaluation Approach, Paper No.10, Occasional Papers: Issues in Training, Training Center, IRRI, International Rice Research Institute.
6. Phillips, J. and Stone, R. D. (2002), How to measure training results: A practical guide to tracking the six key indicators, McGraw-Hill, New York. Patton, M.Q. (1997), Utilization-focused E.
7. Patton, M.Q. (1997), Utilization-focused Evaluation: The New Century Text, Thousand Oaks, CA: Sage Publishers.
8. Reay, D.G. (1994), Evaluating Training, Kogan Page, London.
9. Saxena, R.N. (2006), "Impact Evaluation of Personal Growth Training: A Quasi-experimental Design", ASCI Journal of Management, Vol. 35(1 and 2), March 2006.
10. Tracey, William R (1968), Evaluating Training and Development Systems, American Management Association, New York.

Suggested Reading

1. ADB. 2009. Training Needs Assessment and Strategic Training Plan.
2. Bentaya GM, and Hoffmann V (Eds). 2011. Rural Extension Volume 3-Training Concepts and Tools.
3. Margraf Publishers GmbH, Scientific books, KanalstraBe 21; D-97990, Weikersheim, 191 pp
4. CIDA (1997), Guide to Gender Sensitive Indicators, Canada
5. DFID.2003. Promoting Institutional and Organisational Development. A Source Book of Tools and Techniques, Department for International Development, United Kingdom
6. Development Policy, January 1998, United Nations Development Programme
7. DoPT.2014. Civil Services Competency Dictionary: Strengthening Human Resource Management of Civil Service. Department of Personnel and Training, Government of India
8. FAO.2010. FAO Capacity Assessment Approach and Supporting Tools - Discussion Draft, Food and Agriculture Organisation of the United Nations
9. FAO.2012. Capacity Development: Learning Module 2. FAO Approaches to Capacity Development in Programming. Processes and Tools, Food and Agriculture Organisation of the United Nations
10. FAO.2012. Corporate Strategy on Capacity Development.
11. FAO.2013. Capacity Development: Learning Module 4. Organization Analysis and Development Food and Agriculture Organisation of the United Nations
12. Hanna, G. S., and Dettmer, P. A. (2004). *Assessment for effective teaching: Using context-adaptive planning.* Boston, MA: Pearson A and B
13. GFRAS. 2012. The New Extensionist: Roles, Strategies, and Capacities to Strengthen Extension and Advisory Services, Global Forum for Advisory Services
14. GFRAS. 2015. The New Extensionist: Core Competencies for Individuals, GFRAS Brief 3. Horton D. 2002. Planning, Implementing, and Evaluating Capacity Development. ISNAR Briefing Paper 50.
15. ICAR 2015. Training Policy 2015, Indian Council of Agricultural Research.
16. IISD 2015. Appreciative Inquiry and Community Development. International Institute for Sustainable Development
17. Kirkpatrick, L. D. and Kirkpatrick, D. J. (2005), Evaluating Training Programs–The Four Levels, Benett Koehler Publishers Inc.
18. Laird, D. (1978), Approaches to Training and Development/Reading, Mass, Addison-Wesley Publishing Company

19. LENCD 2011. How to assess existing capacity and define capacity needs, Learning Network on Capacity Development.

20. Maguire. 2012. Module 2: Agricultural Education and Training to Support Agricultural Innovation Systems. Overview. Agricultural Innovation Systems: An Investment Source book. The World Bank.

21. Marcotte, P. L., Bakker D. R. and Bell, M. (2002), Assessing Training Impact: IRRI's New Evaluation Approach, Paper No.10, Occasional Papers: Issues in Training, Training Center, IRRI, International Rice Research Institute

22. Mbabu A N and Hall A. 2012. Capacity Building for Agricultural Research For Development Lessons from Practice in Papua New Guinea. United Nations University-Maastricht Economic and Social Research Institute on Innovation and Technology (UNU-MERIT).

23. Mittal N, Sulaiman RV and Prasad R M. 2016. Assessing Capacity Needs of Extension and Advisory Services a Guide for Facilitators. Agricultural Extension in South Asia. http://www.aesanetwork.org/assessing-capacity-needs-of-extension-and-advisory-services-aguide-for-facilitators/

24. Mishra DC. 1990. New Directions in Extension Training. Directorate of Extension, Ministry of Agriculture, Govt. of India, New Delhi.

25. OECD/DAC. 2006. The Challenge of Capacity Development: Working Towards Good Practice, Organisation for Economic Cooperation and Development.

26. Patton, M.Q. (1997), Utilization-focused Evaluation: The New Century Text, Thousand Oaks, CA: Sage Publishers

27. Phillips, J. and Stone, R. D. (2002), How to measure training results: A practical guide to tracking the six key indicators, McGraw-Hill, New York. Patton, M.Q. (1997), Utilization-focused E

28. Pretty JN, Gujit I, Thompson J, and Scoones I. 1995. A Trainer's Guide for Participatory Learning and Action. IEED Participatory Methodology Series.

29. Reay, D.G. (1994), Evaluating Training, Kogan Page, London

30. Rolf PL and Udai P. 1992. Facilitating Development: Readings for Trainers, Consultants and Policy-makers, New Delhi: Sage Publications, pp. 359

31. Rolf PL and Udai P. 1990. Training for Development, (3rd edn) by (West Hartford, Kumarian Press, 1990, pp. 333.

32. Saxena, R.N. (2006), "Impact Evaluation of Personal Growth Training: A Quasi-experimental Design", ASCI Journal of Management, Vol. 35(1 and 2), March 2006

33. SIDA.2000. Capacity Development. SIDA Working Paper No. 4. Analysis of Needs for Capacity Development.

34. SIDA. 2000. Working Paper No. 4. Analysis of Needs for Capacity Development

35. Sulaiman RV and Mittal N. 2016. Capacity Needs of Extension and Advisory Services (EAS) in South Asia. Policy Brief No 1. Agricultural Extension in South Asia. http://www.aesanetwork.org/policy-brief-no-1-capacity-needs-of-extension-and-advisory-serviceseas-in-south-asia/
36. Swanson BE and Rajalahti R. 2010. Strengthening Agricultural Extension and Advisory Services. A Guide for Facilitators.
37. TAP. 2013. Capacity Development for Agricultural Innovation Systems - Key Concepts and Definitions. Tropical Agricultural Platform
38. TAP. 2016. Common Framework on Capacity Development for Agricultural Innovation Systems. Guidance Note on Operationalization, Tropical Agricultural Platform
39. UNDP. 1998. Capacity Assessment and Development in a Systems and Strategic Management Context. Technical Advisory Paper No. Management Development and Governance Division Bureau
40. Development Policy, January 1998, United Nations Development Programme
41. UNDP. 1998. Capacity Assessment and Development in a Systems and Strategic Management Context. Technical Advisory UNU-MERIT, Netherlands.
42. UNDP. 2008. Capacity Assessment Methodology. User's Guide. Capacity Development Group. Bureau for Development Policy.
43. UNDP. 2009. Capacity Development: A UNDP Primer, United Nations Development Programme
44. WAC. 2013. Assessing Capacity Needs and Strategy Development for Grassroots Rural Institutions: A Guide for Facilitators. World Agroforestry Centre (WAC)

Index